# International Scientific and Research Conference on Topical Issues in Aeronautics and Astronautics 2015



Dedicated to the 55th Anniversary from the Foundation of SibSAU

IOP Conference Series: Materials Science and Engineering Volume 94

Flic-en-Flac, Mauritius
6 - 10 April 2015

ISBN: 978-1-5108-1465-3
ISSN: 1757-8981

**Printed from e-media with permission by:**

Curran Associates, Inc.
57 Morehouse Lane
Red Hook, NY 12571

**Some format issues inherent in the e-media version may also appear in this print version.**



Printed by Curran Associates, Inc. (2016)

For permission requests, please contact the Institute of Physics at the address below.

Institute of Physics
Dirac House, Temple Back
Bristol BS1 6BE UK

Phone: 44 1 17 929 7481
Fax: 44 1 17 920 0979

techtracking@iop.org

**Additional copies of this publication are available from:**

Curran Associates, Inc.
57 Morehouse Lane
Red Hook, NY 12571 USA
Phone: 845-758-0400
Fax: 845-758-2634
Email: curran@proceedings.com
Web: www.proceedings.com

# Table of contents

## Volume 94

**International Scientific and Research Conference on Topical Issues in Aeronautics and Astronautics (dedicated to the 55th anniversary from the foundation of SibSAU), 6–10 April 2015, Krasnoyarsk, Russia**


## Accepted papers received: 22 September 2015
## Published online:


## Preface



## Papers

# International Scientific and Research Conference "Topical Issues in Aeronautics and Astronautics"

## dedicated to the 55th anniversary from the foundation of SibSAU

The International Scientific and Research Conference "Topical Issues in Aeronautics and Astronautics" is one of the most significant scientific conferences arranged by the Reshetnev Siberian State Aerospace University (SibSAU) which is located in the Krasnoyarsk Region of Russian Federation. In April 2015 this Conference was dedicated to the 55th anniversary from the foundation of the University.

Traditionally, the Conference is seen as emblematic of the University's specialty and is annually organized in April, when the first human travelled into space.

This Conference is arranged for undergraduate, graduate and postgraduate students, scientists and lecturers, as well as developers, designers and constructors representing leading companies and enterprises of the aerospace sector to give opportunities to present their projects, research work and results. The Conference is a great chance to connect scientists and highly-qualified and skilled specialists with a new community of future scientists and practitioners in the aerospace sector.

The Conference proceedings include papers presented by creative young specialists closely connected with aviation and space vehicles – design, production, problem-solving in space machine building and aerospace education, macro- and microeconomic development of the field, new approaches to solving philosophical and social problems, – experienced scientists and specialists, and all those who want to dedicate themselves to aeronautics and astronautics.

The selected papers are presented in these proceedings to share University research results, innovations and cutting-edge technologies with the international community to develop aeronautics and astronautics on a global scale.

IOP Conf. Series: Materials Science and Engineering **94** (2015) 011002 doi:10.1088/1757-899X/94/1/011002

# Peer review statement

All papers published in this volume of *IOP Conference Series: Materials Science and Engineering* have been peer reviewed through processes administered by the proceedings Editors. Reviews were conducted by expert referees to the professional and scientific standards expected of a proceedings journal published by IOP Publishing.

# MATHEMATICAL FORMULATION OF TECHNOLOGICAL PROCESSES OPTIMIZATION PROBLEM*

**A A Stupina[1], A A Shigina[2], A O Shigin[3]**
[1] Professor, Department of System Analysis and Operation Research,
Siberian State Aerospace University named after academician M. F. Reshetnev, Krasnoyarsk, Russia
Doctor of tech. sciences, Professor, Head of "Economics and information technologies of management" Department,
Siberian Federal University, Krasnoyarsk, Russia
[2] Post-graduate student of "Economics and information technologies of management" Department
Siberian Federal University, Krasnoyarsk, Russia
[3] Candidate of tech. sciences, Associate professor of "Mining machines and complexes" Department,
Siberian Federal University, Krasnoyarsk, Russia

E-mail: saa55@rambler.ru
shigina_a@mail.ru
shigin27@rambler.ru.

**Abstract.** In this article the methodological principles of creation of adequate mathematical models of the technological processes, which are focused on the use in the tasks of optimization. In this article the task multiobjective parameter optimization formulates at with several control variables in the conditions of uncertainty is considered. Minimizing measurement errors requires introduction to technical system of an adaptive element, which allows it to consider some uncertainty during modeling and optimization.

The complexity of the criteria optimization selection or performance indicator (objective function) consists of the fact that in practice, optimization and management problems deal with a lot of criteria, that are often mutually contradictory. In private problems of optimization when the object is a part of technological process, it isn't always possible or it isn't always expedient to allocate a direct economic indicator, which would completely characterize the efficiency of the investigated object. In such cases, as the criterion of optimality can serve technical characteristics, indirectly assessing the efficiency of work of technical object [1].

Now a difficult and insufficiently studied problem in the solution of optimization tasks exists – the need to achieve a large number of objectives, which leads to emergence of a multicriteria situation in the process of decision-making. Besides, when growth of complexity and dimension of a model take place, big problems with sustainability of optimizing tasks appear. For solving optimization problem it is necessary extremely correct formulation of optimization tasks and development of algorithms for its solution, which should be based on

* Results were obtained in the framework of the state task № 346 of the Ministry of Education and Science of the Russian Federation

the available for modern level of development information.

Usually in all tasks of calculation and optimization methods of sustainable approximate solution of incorrectly posed problems, based on using the additional a priori information about the desired solution are used. The application of these methods is caused by the fact that the data for the calculation obtained from sensors and instrumentations always have considerable error [2]. Even minor fluctuations in the parameters can cause to disruptions in the modes of operation of a technical object. For minimizing the measurement error is necessary to introduce a technical system of the adaptive element [3], allowing it to take into account the uncertainties in the modeling and optimization, respond to changes in the properties of the object of influence on time and maintain the functioning parameters of the control object in optimal proportion [4]. When getting achievement of requirements to output parameters of the process the statistical analysis of the results of system implementation is carried out [5].

It is necessary to apply the mixed criteria representing functions from primary parameters (a multi-purpose problem of management) to a mathematical problem definition of optimization of the studied technological process. As stochastic external influences constantly have impact on the studied object of management in the functioning mode and it is necessary to optimize two clashing criterion functions in the set range of definition, then the task of multiobjective (vector) parameter optimization with discrete- continuous character with several controlling variables in the conditions of uncertainty is set. This task is set for the purpose of optimum process control during the unsteady modes of operation.

For application of quantitative methods for solving optimization problems it is necessary to have mathematical models of optimization criterion and restrictions:

$$W = W(x, a, y), \tag{1}$$

where $x$ – a vector of external conditions or variables on which it is impossible to influence;

$a$ – constant for this task parameters;

$y$ – a control vector or controllable factors the search of value of which is the optimization purpose.

The most effective methods in the terms of uncertainty are those based on a priori model of drilling and searches of an extremum optimization. On the basis of the chosen optimality criterion the compiled objective function representing the dependence of the optimality criterion on parameters influencing its value is made. So, the mathematical formulation of optimization problems of the studied technological process comes down to the determination of extremum (maximum or minimum) of the function (optimization criterion) one or more variables, under certain conditions, which are imposed on some other variables, so this optimization will be conditional.

For the solution of the problem of technological process optimization it is necessary:

1. To choose the optimization criterion.
2. To choose the variable parameters.
3. To develop the mathematical model of technological process.
   3.1. To construct an objective function that specifies connection between the criterion and the variable parameters.
   3.2. To define the system of restrictions imposed on variable parameters.
4. To select the methods of calculation and optimization of a mathematical model.
5. To realize optimization process.

6. To analyze the results (if results are unsatisfactory, return on stages 1, 2, 3.2).

The complexity of drilling processes and the stochastic nature of external influences extremely complicates the work on optimization. During the drilling of exploratory wells for solid minerals optimization of drilling process is the most difficult due to the variability of conditions for drilling operations, insufficient amount of information about the physicomechanical properties of rocks. The complexity of the selection criterion of optimization of regime parameters of drilling consists in the difficulty of its operational calculations that is caused by insufficient necessary information which is continuously supplied from the sensors to the operator of the drilling rig from the dashboard.

Simulation of dynamic processes of functioning of the drilling rig is carried out to optimize the parameters of the drilling process [3, 6]. Development of a technique calculating the parameters of the mathematical models of the drilling process can significantly reduce the time of definition of static and dynamic characteristics of the object, assess the area of their application, reduce the time of adjustment, increase the accuracy of definition and maintenance of optimum regime parameters. So, the efficiency of functioning of all system increases [7]. The complex solution of the tasks connected with creation of a mathematical model, collection and processing systems of primary information and optimization the mode of drilling, is a big scientific problem, which is of great importance for the national economy. The feature of optimization problems is that the variables in the optimization criterion and restrictions often have discrete nature (eg, the number of drill bits).

The aspiration to optimize the drilling process has led to attempts of models of this process creating, which are based on the theoretical calculations and empirical regularities that characterize the working capacity of drill bits and the efficiency of rock destruction. In domestic practice models of the drilling process were formed after the establishment of dependence of rate of penetration $V$ from bit load $G$ and the frequency of rotation n made by V.S. Fedorov [8], obtained empirical dependence:

$$V = an^x G^y, \tag{2}$$

where $x$, $y$, $a$ – the correction factors.

As criteria that need to be optimized, it is proposed to choose: drilling speed, resource (tunneling) - observed during the drilling parameters; the reliability of the bit, capacity of feeders and rotation, energy consumption (energy intensity of drilling) and the cost of drilling of one meter of hole - not observable parameters, but quickly computable. Such criteria as performance, speed of drilling, resource, reliability of the bit, profits, it is necessary to maximize. It is necessary to establish a permanent monitoring of the amplitude and frequency parameters of dynamical loads. As an integral indicator, which brings together all these criteria it is offered to introduce a specific drilling cost (rub. / linear meters). The proposed optimization criteria can be used as the basis for the development of control algorithms roller-cone machine in drilling mode.

The decision-making problem definition, understood as a problem of selecting the optimum drilling mode, can be formed in the following way: suppose $Y$ - a set of alternatives (drilling mode options); $X$ - a set of the parameters characterizing the drilling mode; $Z$ - set of possible outcomes (costs); $X$, $Y$, $Z$ - any abstract sets. Existence of a causal relationship between a choice of some alternative of $y_i \in Y$ and approach of some outcome of $z_i \in Z$ is supposed. The outcome of $z_i \in Z$ can be estimated by a real number according to some map: $f\colon Z \to R$, in our case it is well-drilling cost.

The problem of optimizing the drilling process comes down to finding the extremum of the objective function:

$$R = f(u) = extr, \tag{3}$$

where $R$ - multidimensional function.

Provided that the input control actions for the drilling machine are corrected values of the axial force, frequency of rotation, power, the amount of air to clean the well, the expression (3) will be as follows:

$$R = f(P_{cor}, n_{cor}, N_{cor}, Q_{cor}) = extr, \tag{4}$$

where $P_{cor}$ – the corrected value of axial force, $P_{cor}$ = $=P + \Delta P$;
$n_{cor}$ – the corrected value of frequency of rotation, $n_{cor}$ = $=n+\Delta n$;
$N_{cor}$ – the corrected power value of feeders, $N_{cor} = N+\Delta N$;
$Q_{cor}$ – the adjusted value of amount of air for a hole cleanup, $Q_{cor} = Q + \Delta Q$.

Here $\Delta P$, $\Delta n$, $\Delta N$, $\Delta Q$ – correction values of axial force, frequency of rotation, power of feeders and the amount of air. These values are obtained in the controller by using calculation methods that assess drilling tool resource and economic efficiency of the drilling process [7] and the procedure for determining of settlement firmness of drill bits with complex loads, depending on the characteristics of the breed and drilling modes [9].

The process of making decisions on the choice of the optimal drilling mode in the classical systems theory has two goals:

- the purpose of the restriction $z_0 \leq z_i(t) \leq z_1$, $i = \overline{1,n}$ – demands finding of target variables in such defined borders, with which the condition of the system is stable or optimum $y(x) \to opt$, where $x \in X$ – any set of the parameters describing a condition of system. In this case the restrictions are strength range [10] and cyclic firmness of a chisel, a resource and the maximum capacity of a drilling rig, the minimum speed of drilling.
- extreme purpose is to maintain the target variables in extreme conditions $y_i(x) \to extr$, $i = \overline{1,n}$. This purpose consists in productivity maximizing with the set prime cost (criterion of optimization - productivity) and the minimization of prime cost with the set productivity (criterion of optimization - prime cost).

In this way, the general problem definition of decision-making can be formulated as follows: to find $X_{1opt}$, $X_{2opt}$, ..., $X_{iopt}$, belonging to the areas of their admissible values $W_{X1}, W_{x2}, ..., W_{Xi}$, which whenever possible will turn into a maximum (minimum) the criterion of an optimality Z.

The problem definition of multiobjective optimization of regime parameters for the studied technological process is as follows:

$$z = \{y_1(\bar{x}), y_2(\bar{x}), y_i(\bar{x}), ..., y_k(\bar{x})\} \to extr, i = \overline{1,k}, \tag{5}$$

where $y_i : R^n \to R$ – it is $k$ of objective functions, ($k \geq 2$) (productivity, the cost of drilling, load, etc.).

Vectors of decisions $\bar{x} = (x_1, x_2, ..., x_n)^{\tau}$ belong to non-empty range of definition $\bar{x} \in W$.

Each of local criteria is characterized by the certain coefficient of relative importance. We will designate these coefficients as $\lambda_1$, $\lambda_2$, ...,$\lambda_k$. As well as the set of local or private criteria form integrated or vector criterion of an optimality $Z=\{Z_k\}$, coefficients of relative importance $\lambda_i$ form the importance vector: $\overline{\lambda}=\{\lambda_k\}$.

The task consists in finding the optimum value of X from the area of admissible values $W_X$. Each local criterion characterizes one quality of the decision being made. The optimal decision X can be written conditionally as follows:

$$F=\overline{F}(\overline{X},\overline{Y})=\underset{X\in W_X}{opt}\left[\overline{Z}(\overline{X},\overline{Y}),\overline{\lambda}\right], \quad (6)$$

where $F$ – the optimal value of the integral criterion;

$X$ – optimum values of the operated parameters;

*opt* – the operator of optimization which defines the selected optimization principle;

$\overline{\lambda}$ – vector of importance of local criteria.

The area of admissible values can be devided into two not crossed subareas: $W_X^C$ – the area of "consent" in which the quality of the made decision can be improved by one or several local criteria without deterioration at least of one of the remained local criteria; $W_X^K$ – the area of "compromises" in which the improvement of the decision on one or several local criteria surely leads to the decrease in values of one or several remained local criteria.

By means of an integral indicator optimization is possible for a sufficiently long period of time within which to carry out several complete cycles process, analyze the results and to find the optimum conditions of the system functioning. Found the extremums of the objective function in the optimization problem of the drilling process allow us to determine the best values of the parameters of the investigated process in conditions of indeterminacy. Corrected values obtained by using computational techniques that can give a integrated assessment of the drilling process economic efficiency. The adaptive element in the system and methods based on a priori model of the technological process and the search for the optimization extremum is necessary for precision of measurements and uncertainty in modeling and optimization.

Summing up the results, it should be noted that the criterion of optimization - characteristic attribute of the solution of a task, according to the value of which the optimality of the found solution is estimated. In the studied problem of optimization of technological process some criteria of an optimality should be established. It is necessary to set a task of multiobjective parameter optimization with the purpose of optimum control of technological process with the unsteady modes of operation. The methods based on a priori model of technological process and searches of an extremum of optimization are the most effective in the conditions of uncertainty.

## References

[1] Nazarian S.A. 2004 Development of methods of reliability management of complex technical systems with dependent failures of elements (doctoral thesis): Lipetsk. 140 p.

[2] Tikhonov A.N. *et al 1983* Regularizing algorithms and a priori information. Moscow: Nauka. 200 p.

[3] Stupina A.A., Shigina A.A., Shigin A.O., Karaseva M.V. and Ezhemanskaja S. N. 2014 Automated intellectual system with the short-duration nature of feedback. *Life Sci J*. Vol. 11, № 8s. pp. 302-306.

[4] Stupina A.A., Shigin A.O., Shigina A.A., Karaseva M.V. and Korpacheva L.N. 2014

Control and Management by Resource of Rolling Cutter Bits in Drilling Rock Massif. *Middle-East J. Sci. Res.* Vol. 21, № 1. pp. 84-90.
[5] Ivchenko B.P., Martyshenko L.A. and Monastyrski M.L. 1997 The theoretical foundations of informational and statistical analysis of complex systems. St. Petersburg: Lan. 320 p.
[6] Shigina A.A., Shigin A.O. and Stupina A.A. 2012 Comparative evaluation of methods for analyzing the efficiency of the drilling rigs. *Modern problems of science and education* **6** 11p.
[7] Stupina A.A., Shigina A.A. and Shigin A.O. 2013 The analysis of efficiency of the multi-parameter system *Vestnik the Siberian State Aerospace University named by Academician M.F. Reshetnev* **2 (48)** pp. 94-100.
[8] Fedorov V.S. 1958 Engineering of drilling modes. Moscow: Gostoptekhizdat. 215 p.
[9] Shigin A.O., Gilyov A.V. and Shigina A.A. 2013 Voltage and resistance roller cone bits in drilling complex structural rock massifs *Mining informational and analytical bulletin* **4** pp. 325-333.
[10] Gilyov A.V., Shigin A.O. *et al* 2011 Methods of strength calculations for the engineering of working parts drilling rigs *Modern knowledge-intensive technologies* **1** 3 p.

IOP Conf. Series: Materials Science and Engineering **94** (2015) 012002 doi:10.1088/1757-899X/94/1/012002

# SPACEWIRE ROUTER IP-CORE WITH PRIORITY ADAPTIVE ROUTING*

**A V Shakhmatov[1], S A Chekmarev[1], M Y Vergasov[1], V Kh Khanov[2]**
Siberian State Aerospace University Named After Academician
M.F. Reshetnev (SibSAU)
31 "Krasnoyarskiy Rabochiy" Prospect, Krasnoyarsk, 660037, Russia.
[1]Engineers Of SibSAU
[2] Ph.D., Assistant Professor Of SibSAU

E-mail: khvkh@mail.ru

**Abstract.** Design of modern spacecraft focuses on using network principles of interaction on-board equipment, in particular in network SpaceWire. Routers are an integral part of most SpaceWire networks. The paper presents an adaptive routing algorithm with a prioritization, allowing more flexibility to manage the routing process. This algorithm is designed to transmit SpaceWire packets over a redundant network. Also a method is proposed for rapid restoration of working capacity after power by saving the routing table and the router configuration in an external non-volatile memory. The proposed solutions used to create IP-core router, and then tested in the FPGA device. The results illustrate the realizability and rationality of the proposed solutions.

## 1. Introduction

Providing network communication of onboard equipment is a leading direction of Space Instrument development. The most important network technology for space applications is SpaceWire, widespread use of which has led to an emergence of many varieties of different network infrastructure devices such as routers, including in the form of IP-cores.

Lately kinds of router architectures and its functionality settled and generally the same for all providers. SpaceWire standard, in the main, does not allow of variation of the internal routers architecture. Differences are the types of external interfaces for accession to the host processors. The works [1-4] present examples of implemented on the basis of IP-core routers from several providers. Nevertheless, several important practical issues are not reflected in the architecture of routers. For example, the routers redundant network structure, when alternative SpaceWire channels are created to improve reliability and overall network throughput. Also an important attribute for a large redundant network is the ability of autonomous to preserve the configuration and routing tables when power is off of the router. Power may be disabled emergency or scheduled but in both cases is often important recovery rate of the router performance.

A task of this paper is to present solutions to these problems. The first problem is carried out with a help of a group adaptive routing, using prioritizations of router output ports; the second with the help of an external non-volatile memory of the router, which stores the current configuration and routing table.

*The research was done with the financial support of the Ministry of Education and Science of the Russian Federation in accordance with the agreement № 14.574.21.0041, unique identifier RFMEFI57414X0041

These decisions are reflected in the developed reconfigurable VHDL IP-core of router. On the basis of the IP-core were created stand-alone router, and the router built-in system-on-chip. The results confirmed rationality of the developed solutions.

## 2. Known solutions

Reliability of the equipment is a prerequisite for space missions. Reliability of equipment is provided of hardware redundancy and reliability of the network is further provided redundancy of the paths. Fault tolerant architecture provides redundant paths for routing of packets.

Paragraph [5] show diagrams of the fault-tolerant networks when the core network are duplicated with its infrastructure components (routers) and nodes i.e. have the same primary and backup parts of the sub-networks. Nodes of main and redundant networks are linked with main and redundant routers. Fault tolerance is provided with the redundancy SpaceWire packages sent.

Section [6] demonstrate a fault tolerant topology scheme of a double star, where each node is considered as dual redundant, i.e. primary and redundant. The nodes can be hot redundant or cold redundant. It is proposed to use a special distributed fault tolerant routing protocol to support fault tolerant routing topology in [6]. To perform fault tolerant routing protocol all network devices should have the intelligent capability which makes network hardware more complex.

The work [7] consider several possible fault tolerant routing topologies, based on known types of network structures such as a double star, a double-double star, a triple star and similar. At least for small networks, e.g. for small spacecrafts, a double star are considered as most relevant solution. As well as [6] it is proposed to use one embodiment of a distributed adaptive routing algorithm. However, for the consideration of the small size of the network adaptive centralized routing [8] would be more appropriate.

In the above works fault tolerant topologies are supported by redundant network traffic. However, certain standard [9] group adaptive routing also can be used for reserving data channels. To do this, it must be supplemented prioritization output ports, which allows for the logical address of the packet arrived at the router is easy to determine the output port according to its priority and the lowest priority to reserve the output port corresponding SpaceWire channel. In this case, there is no redundancy of network traffic via additional redundant packets or routing protocol packets. This paper presents the rationale of the proposed solutions and supports its implementation in a router.

In addition, the known solutions sidestep the issue of minimizing delays recovery routing table and router configuration during power loss. Turning off the power of the router is very likely situation. It is assumed that the restoration of the router settings should provide the host computer. However, if the network is large, besides redundant, then the recovery process settings can be time consuming, that in some cases is unacceptable. The paper proposes to save the settings of the router in the external memory. When the router is turned on, its settings are loaded from the external memory, and network immediately becomes operational.

These solutions have been implemented in developed reconfigurable VHDL IP-core router. Four-port stand-alone router was created on the basis of IP-core. The results confirmed the effectiveness of the developed solutions.

## 3. Proposed solutions

### *3.1. Group adaptive routing using prioritization output port*

Multiple output ports of the router to one logical address of the packet, received in the router, conform to redundancy channels. Group adaptive routing provides conformity of single logical address of group output ports. Selecting a particular output port of the group allows prioritization of ports.

Suppose one logical address corresponds to the four output ports. Each output port, can specify multiple priority levels, for example 4. When a packet arrives with this logical address it is sent to the port with the highest priority. If the port with the highest priority is busy, the packet is routed to the port with the lower priority level. If it is busy too, the packet is routed to the port with much lower priority level. If several free ports have the same priority level, the packet crossing in the port with the lowest port number.

The use of priorities is easiest to illustrate via tables. Fig. 1 shows the routing and priorities tables for the router, which has 4 ports.

| | Address | Port 0 | Port 1 | Port 2 | Port 3 | Port 4 |
|---|---|---|---|---|---|---|
| Routing | 0 | 1 | 0 | 0 | 0 | 0 |
| Path Addressing | 1 | 0 | 1 | 0 | 0 | 0 |
| | 2 | 0 | 0 | 1 | 0 | 0 |
| | ... | | | | | |
| Logical Addressing | 32 | 0 | 1 | 1 | 0 | 0 |
| | 33 | 0 | 0 | 0 | 1 | 1 |
| | 34 | 0 | 1 | 1 | 0 | 1 |
| | ... | | | | | |
| | 255 | 0 | 0 | 0 | 0 | 0 |

| | Address | Port 0 | Port 1 | Port 2 | Port 3 | Port 4 |
|---|---|---|---|---|---|---|
| Prioritization | 0 | x | x | x | x | x |
| Path Addressing | 1 | x | x | x | x | x |
| | 2 | x | x | x | x | x |
| | ... | | | | | |
| Logical Addressing | 32 | x | 00 | 11 | x | x |
| | 33 | x | x | x | 10 | 11 |
| | 34 | x | 11 | 10 | x | 01 |
| | ... | | | | | |
| | 255 | x | x | x | x | x |

**Figure 1.** Table of routes and priorities.

Logical address 32 has two output ports: port 1 and port 2. When receiving a packet with this address, it is routed to port 2 having a higher priority. If port 2 is busy, the packet will be forwarded to port 1 which has less priority. If the logical address has only 2 ports, it is enough only 2 priority levels: low and high. If the network is more redundant, the logical address may have more than 2 output ports, it is necessary to increase the number of priority levels. So the logical address of 34 has three output ports and each of them may have, in the general case, own priority. For most practical cases it is sufficient to have four levels of priority.

Multi-level prioritization can be used to prohibit using of certain ports without disconnecting them in a router. For example, you can set some ports the lowest priority level and prohibit their use. The packages will be sent to the ports with higher priority, and if these ports are busy, the packets will wait until they become free. Subsequently, if one of the channels connected to a port with high priority will refuse, ban on the use of the port with the lowest priority will be removed. Thus, actually made a reservation channels and manage their switching via group adaptive routing with prioritization of the ports. The method is very simple and easy to implement in the router.

Should be noted another point which previously discussed in part [8, p.99]. If the router simultaneously received two packages with different logical addresses for which the highest priority has the same port, to avoid conflict first access to the table will receive a package whose logical address is smaller. It takes the port with a higher priority, and the packet with a lower logical address will be sent to the port with lower priority.

Note also that to implement the storage of priorities easier just not as one table but as two tables, each of which stores only 1 bit of 2-bit representation of the priority level. Although other options are possible.

Thus, the algorithm group adaptive routing with prioritization of ports produces: backup of paths, switching to a backup paths, sending packets over paths with higher priority.

### *3.2. External backup router configuration*

When the router is turned off and then the power is restored immediately only path routing is available. In large networks with alternative data channels using path routing is impractical. The host computer performs recovery of logical addresses in the routing tables, a content of priority table and the configuration of the router. Redundant network has at least twice the number of routers. Therefore, this process can take a long time, during which the routers and the network as a whole inoperative. To minimize the recovery time of work is offered each router to equip with an external non-volatile memory. In this memory stores all the settings of the router that when power is restored very quickly loaded into the router, which almost immediately becomes ready for operation. The network will become operational faster than after you disconnect and restore power will boot of the host computer.

## 4. An example implementation

The presented solutions have been implemented in reconfigurable router IP-core created in VHDL. Reconfiguration refers to the ability of choosing from the total possible number of ports required number of SpaceWire ports and FIFO interfaces synthesized for a specific router. Maximum number of ports SpaceWire ports and FIFO interfaces is 31. The rest of the structure is quite traditional router (Figure 2) with the exception of the components responsible for the implementation of the proposed solutions. Port priorities are stored in the priority tables. Request to the external memory which stores the configuration of the router by using a flash ROM controller via SPI interface.

All ports are connected to each other by using non-blocking switching matrix, which allows establishing a direct channel between them. And already established connections do not prevent the establishment of new, if switching ports free.

Using of regional-logical addressing involves the need to remove the first byte of transmitted package. For this purpose is using the register, in which sets the number of regional ports.

Additionally SpaceWire router supports broadcast distribution of packages over network. This function can be set for one or more logical addresses. Upon arrival of the package it will be sent to all the ports listed in the routing table for that logical address.

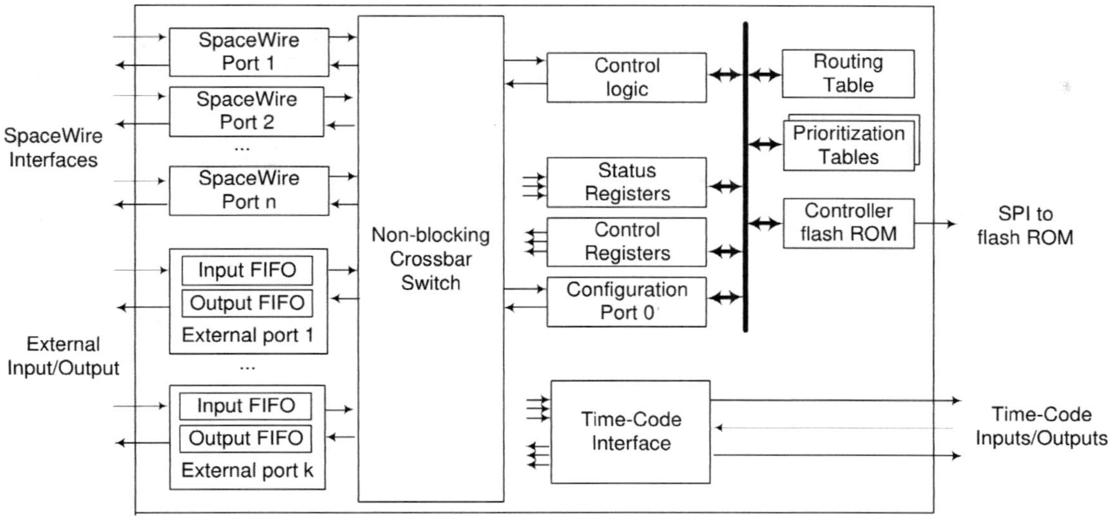


**Figure 2.** SpaceWire Router Block Diagram.

For configuration uses an internal zero port that supports the RMAP protocol in accordance with ECSS-E-ST-50-52C standard. RMAP-target allows you to read and write data directly into the registers of the router.

The router allows you to collect status information about a number of errors associated with errors routing of packet, a timeout of the port, breaks SpaceWire connection. Additionally, each port has its own status register in which is recording the detailed information about its condition.

Functional verification of the router, as well as RMAP-target, was carried out by creating full-featured models of testing environment in FPGA. The models consisted of a system-on-chip that include the device under test and one or more SpaceWire codec [9, 10]. These network devices communicate via external cables forming SpaceWire channels created in this way fragment of the network. Test software running on processor core system-on-chip, carried out exchange of a test packets, via which was verified of a router function.

## 5. Experimental results

For evaluation of FPGA resources which are required to accommodate the designed IP-core the router was synthesized in several configurations. As the FPGA Altera EP4CE115F29C7, Microsemi A3PE1500 and Microsemi RTAX2000 were used. The results of synthesis are shown in table 1. They can be considered satisfactory.

**Table 1.** Results of the synthesis of router in FPGAs.

| Number of ports | Altera EPCE115F | | Microsemi A3PE1500 | | Microsemi RTAX2000 | |
|---|---|---|---|---|---|---|
| | Les,% | FFs,% | Cells,% | FFs,% | Cells,% | FFs,% |
| 4 | 3 | <1 | 32 | 15 | 24 | 13 |
| 8 | 6 | <1 | 48 | 28 | 43 | 26 |
| 16 | 12 | <1 | 91 | 55 | 78 | 51 |
| 20 | 15 | <1 | 114 | 73 | 98 | 68 |

On the basis of the developed IP core have been established 4-port stand-alone router and the router which are built into the system-on-chip based on the LEON3 [10]. Microsemi A3PE1500 and Microsemi A3PE3000 of FPGA were used, respectively. The maximum speed of the network that can be created on the basis of these devices is 200 Mbps. After remove and restore a power the router became fully operational in less than 1 ms.

**6. Conclusion**

This paper presents a priority group adaptive routing algorithm, which allows more flexibility to manage the routing process. It is designed for transmission of SpaceWire packets over a redundant network. The proposed solution is to manage the ports of the router via using and configuring of priorities tables. The solution allows to reserve the channels, switching to the backup channels and sending packets on channels with higher priority. Furthermore, is proposed a method for rapid restore functionality the router after power off using save the routing table and the router configuration in an external non-volatile memory.

The proposed solutions was used to create the router IP-core, and then tested in the FPGA when were creating 4-port stand-alone router and router built into the system-on-chip. The results illustrate the realizability and rationality of the proposed approach.

**References.**

[1] Parkes S, McClements C, Kempf G, Gleiss C, Fischer S, Fabry P 2008 SpW-10X Router ASIC. Testing and Performance *Proceedings of the 2rd International SpaceWire Conference* pp.87-91.

[2]. Isomäki M, Habinc S, Gaisler J 2010 A Configurable SpaceWire Router VHDL IP Core *Proceedings of the 3rd International SpaceWire Conference* pp.229-232.

[3] Isomäki M, Habinc S 2011 Development of a novel 18x SpaceWire router *Proceedings of the 4rd International SpaceWire Conference* pp.293-299.

[4] STAR-Dundee, "SpaceWire Router Mk2S", available at: http://www.stardundee.com/products/spacewire-router-mk2s.

[5] Syschikov A, Suvorova E, Sheynin Y, Sedov B, Matveeva N, Raszhivin D 2013 Toolset for SpaceWire Networks Design and Configuration *Proceedings of the 5th International SpaceWire Conference* pp.149-153.

[6] Fayyaz M, Vladimirova T 2010 Fault Tolerant SpaceWire Routing Topology and Protocol *Proceedings of the 3rd International SpaceWire Conference* pp.407-410.

[7] Nikitin D A, Khanov V Kh, Vergasov M Y, Chekmarev S A 2012 Network architecture of onboard control complex *Proceedings of Conference "Hardware and software control systems, control and measurement"* pp.1539-1546 (In Russ.).

[8] Duato J, Yalamanchili S, Ni L 2003 Interconnection Networks: An Engineering Approach *Elsevier Science* 602p.

[9] ECSS, "SpaceWire - Links, nodes, routers and networks", ECSS-E-ST-50-12C, July 2008, available from http://www.ecss.nl.

[10] Parkes S 2012 SpaceWire User's Guide *STAR-Dundee Limited* 117 p.

[11] Khanov V Kh, Shakhmatov A V, Chekmarev S A 2014 A functional verification system of IP-blocks in network protocols *12$^{TH}$ International Conference On Actual Problems Of Electronic Instrument Engineering Proceedings* (APEIE-2014)) vol.1 pp. 247-251

[12] Khanov V Kh, Shakhmatov A V, Chekmarev S A 2015 A verification system of RMAP protocol controller *IOP Conf. Series: Materials Science and Engineering* **70** 012014.

IOP Conf. Series: Materials Science and Engineering **94** (2015) 012002 doi:10.1088/1757-899X/94/1/012002

[13] Khanov V Kh , Chekmarev S A, Shahmatov V A, Vergasov M Y, Lukin F A 2013 Development of equipment information exchange system of onboard control complex small spacecraft *Vestnik SibGAU* **5(49)** pp. 149-153 (In Russ.).

IOP Conf. Series: Materials Science and Engineering **94** (2015) 012003 doi:10.1088/1757-899X/94/1/012003

# HEURISTICS AND CRITERIA FOR CONSTRUCTING LOGICAL PATTERNS IN DATA*

**A N Antamoshkin[1,2], I S Masich[1] and R I Kuzmich[2]**
[1]Department of System Analysis and Operation Research, Siberian State Aerospace University named after academician M. F. Reshetnev, Krasnoyarsk, Russia
[2]Department of Economics and Information Technology of Management, Siberian Federal University, Krasnoyarsk, Russia

E-mail: i-masich@yandex.ru

**Abstract.** The article considers various optimization models for constructing patterns in the method of logical analysis of data. Application techniques of the proposed models are specified and comparison of their classification against the accuracy on the task of predicting complications of myocardial infarction is provided

## Introduction

The main object of the research conducted by the authors is the method of logical analysis of data derived from the theory of combinatorial optimization [1]. This method was successfully used to solve a number of problems in various areas [2-4]. It belongs to rule based classification algorithms and is based on identification of logical patterns from the original data sample. The original sample consists of two disjoint sets $\Omega^+$ and $\Omega^-$ of observations which are n-dimensional vectors belonging to positive or negative class respectively. Vector elements, also called attributes, can be numerical, nominal or binary. The goal is to determine the class for some new observation, which is also a vector of n variables.

To achieve this goal, a classification model is to be developed, which consists of a set of patterns identified from the original data sample. Each pattern should be informative, i.e. covers a number of objects of the same class and does not cover (or coves a small amount of) the objects of another class [5].

This paper offers several optimization models for constructing a pattern in order to build an adequate classifier able to classify a new incoming object, i.e. an object, previously not participating in its building.

## Model for constructing patterns with maximum coverage

The considered method is used for working with data samples, where attributes take binary values. Since the original sample can consists of different types of attributes, it needs binarizing. In this paper the "unitary" binarization method is used, discussed in detail in [6].

The notion of a pattern [7] is used as the basis for the considered method. A positive pattern can be defined as a subcube of the space of binary variables $B_2^t$, and this subcube intersects with the set $\Omega^+$ and has no common elements with the set $\Omega^-$. A negative pattern is defined similarly.

* Results were obtained in the framework of the state task № 346 of the Ministry of Education and Science of the Russian Federation

The positive ω-pattern for $\omega \in \{0,1\}^t$ is a pattern, which contains ω points. The problem is to find the maximum ω-pattern for each point $\omega \in \Omega^+$, i.e. covering the largest number of points in $\Omega^+$.

Let us define the corresponding subcube using variables $y_j$:

$$y_j = \begin{cases} 1, & \text{if } i-\text{th criterion fixed in the subcube}, \\ 0, & \text{in another case}. \end{cases}$$

That is, by fixing $l$ variables of the original cube with dimension $t$, we get a subcube with dimension $(t-l)$ and with the number of points $2^{t-l}$.

Requirement, which says that a positive pattern should not contain any points in $\Omega^-$, demands that for each observation $\rho \in \Omega^-$ variable $y_j$ takes on a value 1 at least for one $j$, where $\rho_j \neq \omega_j$:

$$\sum_{\substack{j=1 \\ \rho_j \neq \omega_j}}^{t} y_j \geq 1 \text{ for any } \rho \in \Omega^- .$$

For increasing stability to faults, the constraint may be strengthened by replacing the number 1 on the right side of the constraint on the positive integer $d$.

On the other hand, a positive observation of $\sigma \in \Omega^+$ is included in the considered subcube when the variable $y_j$ takes on a value 0 for all indices $j$, where $\sigma_j \neq \omega_j$. Thus, the number of positive observations covered by a ω-pattern can be calculated as:

$$\sum_{\sigma \in \Omega^+} \prod_{\substack{j=1 \\ \sigma_j \neq \omega_j}}^{t} (1 - y_j) .$$

Thus, we have the problem of constrained pseudo-Boolean optimization with algorithmically defined functions:

$$\sum_{\sigma \in \Omega^+} \prod_{\substack{j=1 \\ \sigma_j \neq \omega_j}}^{t} (1 - y_j) \to \max, \tag{1}$$

$$\sum_{\substack{j=1 \\ \rho_j \neq \omega_j}}^{t} y_j \geq d \text{ for any } \rho \in \Omega^- ,\ y \in \{0,1\}^t . \tag{2}$$

The problem of finding the maximum negative pattern is formulated similarly.

Each pattern is characterized by two parameters: coverage, i.e. the number of objects of a particular class, which are covered by this pattern; and degree, i.e. the number of variables involved in its formation. Specificity of recognition problems encountered in practice is in the fact that a database includes a large number of unmeasured values (missing data), and performed measurements may be inaccurate or incorrect. Noises and outliers can lead to the situation where objects of different classes overlap each other, thus entering to the "field" of the opposite class. As a result, obtained patterns have a greater degree and substantially less coverage, than in case of no outliers and inaccuracies. At the same time the classifier consists of a large number of small patterns (with a small coverage). It does not allow constructing an effective classifier with «well-interpreted» conventions (in which a small number of attributes are involved) and with high accuracy recognition.

To increase the stability of the method to outliers, the constraint (2) should be relaxed. It means creating possibilities for a pattern to cover some small number of objects of another class. Then, the degree of obtained patterns reduces and coverage is increased.

Thus, the constraint of the optimization model may be wrighten as follows:

$$\sum_{p\in\Omega^-} z_p \le D, \qquad (3)$$

$$\text{where } z_\rho = \begin{cases} 0, \text{if } \sum\limits_{\substack{j=1 \\ \rho_j \neq \omega_j}}^{t} y_j \ge d, \\ 1, \text{in another case;} \end{cases}$$

$D$ is the number (non-negative integer) of objects of another class which are allowed being covered by the pattern [8].

**The procedure of growing patterns to increase their informativeness**

In accordance to models 1 and 3, the most preferred patterns are those with the greatest coverage. As a result the formed patterns have small degree, i.e. consist of a small number of terms and use only some attributes [8].

Patterns with small degree correspond to large areas in the attribute space. This leads to the possible covering of objects of another class (missing in the training sample) and increasing the number of incorrectly recognized objects. This feature can influence on pattern informativeness reducing it. Therefore, to increase informativeness, we propose the growing procedure for patterns. The growing procedure is applied to each built pattern and means maximizing degree of patterns under fixing coverage:

$$\sum_{j=1}^{t} y_j \to \max,$$

$$fc(Y)=fc'(Y),$$

where $fc(Y)$ is value of the objective function (1) (coverage) for the pattern before the growing procedure, $fc'(Y)$ is value of the objective function for the pattern after the growing procedure.

As a result of the growing procedure, patterns with maximum coverage and with higher degree are formed, thereby increasing reliability of decisions made by the classifier obtained accordance to this heuristic. Reliability of decisions is growing due to increased informativeness of patterns, as the number of objects of "own" class (the value of the objective function for the pattern) remains unchanged, and the number of objects of another class covered by pattern decreases. Thus, the growing procedure contributes to increasing informativeness of each pattern and, consequently, the classifier in general.

**Model for formation of patterns with various covering**

The approach proposed consists in modification of the objective function (1) in order to increase various of patterns in the classifier [9].

The objective function (1) maximizes coverage for each formed pattern, capturing objects that are typical representative of the class, while atypical objects of the class remain uncovered and the classifier has no patterns comprising this objects. Therefore, we obtain a set of similar patterns for the class, thereby reducing the quality of classification. To obtain a classifier with higher various in patterns that allows to select substantially different subsets of objects, it is proposed to modify the objective function (1) for finding the positive pattern as follows:

$$\sum_{\sigma\in\Omega_S^+} K_\sigma \cdot \prod_{\substack{j=1 \\ \sigma_j \neq \omega_j}}^{t} (1-y_j) \to \max, \tag{4}$$

where $K_\sigma$ is weight of positive observation $\sigma\in\Omega_s^+$; this weight should be decreased if this observation is covered by the pattern, thereby this reduce participation priority of the observation in the formation of the next pattern for the sake of uncovered observations.

Similarly, the objective function in the optimization model is formed for finding negative patterns.

In order to use the optimization model with the objective function (4) for the constructing patterns, it is necessary to assign initial weights for all objects and to define rule for weight changes for objects that taken part in formation of the current pattern. The initial weights are proposed to equate to 1 for each object in the training sample. The rule for weight changes for object which took part in the formation of the current pattern:

$$K_{i+1} = \max\left[0,\, K_i - \frac{1}{N_{\max}}\right],$$

where $K_i$, $K_{i+1}$ – weights of the covered object in the formation of the current and next patterns; $N_{\max}$ – parameter defined by the researcher, it means the maximum number of patterns covering the object of the training sample in the classifier.

Thus, using the optimization model with the objective function (4) and the constraint (3) for constructing patterns leads to various patterns obtained.

**Results**

To compare the proposed models of patterns formation, a series of experiments is conducted on the problem of predicting complications of myocardial infarction (MI): ventricular fibrillation (VF), atrial fibrillation (AF), pulmonary edema (PE) [10].

Previously, the problem of predicting complications of MI was solved with use of artificial neural networks [10]. It was observed that the classifier gives decisions which are not good enough in case of a significant difference in the number of objects of each class in the original sample, therefore, the following approach was proposed in that paper to solve this problem. The number of patients with certain complication (positive objects) is approximately ten times less than the number of patients without the complication (negative objects). The original sample (1700 objects) was divided into the test sample and 10 training samples for each complication. Furthermore, the positive objects in training samples remain the same and negative objects change. The classifier was trained at each training sample separately, but it was tested on total test sample. The final decision on each object of the test sample is made by the majority of votes received from all classifiers obtained on the basis of 10 training samples.

For constructing patterns the following four optimization models was used:

1) rigid model that does not allow covering objects of another class by the pattern;

2) relaxed model, which allows the pattern to cover some limited number of objects of another class;

3) modified model including the procedure of growing patterns;

4) model for formation of patterns which covering substantially different subsets of the sample objects.

For finding suboptimal solutions in accordance with optimization models used, heuristic algorithms of constrained pseudo-Boolean optimization described in [11-13] and implemented in the software application [14] was applied.

Classifier forms when combining all built positive and negative patterns.

The following decision rules were used to classify a new (or test) observation [5]:

1) If the observation satisfies the requirements of one or more positive patterns and does not satisfy the requirements of any negative pattern then it is classified as positive.

2) If the observation satisfies the requirements of one or more negative patterns and does not satisfy the requirements of any positive pattern then it is classified as negative.

3) If the observation satisfies the requirements p' from p positive patterns and q' from q negative patterns then sign of observation is defined as $p^{'}/p-q^{'}/q$.

4) If observation does not satisfy the requirements of any patterns, positive or negative, it belongs to class having the lowest price error.

The number of patients with complications (positive objects) and without complications (negative objects) in the training sample and volume of the test sample for all considered complications of MI are presented in table 1.

**Table 1.** Volume of each sample for all complications of MI.

| **Problem** | **Number of positive objects** | **Number of negative objects** | **Number of objects of testing sample** |
|---|---|---|---|
| VF | 70 | 181 | 30 |
| AF | 170 | 180 | 50 |
| PE | 159 | 173 | 39 |

Results of the tests are presented in tables 2-4.

**Table 2.** Classification results for the problem of predicting VF.

| **Optimization problem** | **Set of conditions** | **Number of conditions** | **Covering of negative objects** | **Covering of positive objects** | **Degree of conditions** | **Classification accuracy, %** |
|---|---|---|---|---|---|---|
| The objective function (1), | neg. | 161 | 14 | 0 | 5 | 90 |
| | pos. | 60 | 0 | 14 | 3 | 78 |

| | | | | | | |
|---|---|---|---|---|---|---|
| delimitation (2) | | | | | | |
| The objective function (1), delimitation (3) | neg. | 161 | 26 | 5 | 5 | 90 |
| | pos. | 60 | 5 | 22 | 3 | 78 |
| The objective function (1), delimitation (3) with building procedure | neg. | 161 | 26 | 3 | 6 | 90 |
| | pos. | 60 | 3 | 22 | 4 | 83 |
| The objective function (4), delimitation (3) | neg. | 54 | 16 | 5 | 6 | 80 |
| | pos. | 51 | 5 | 18 | 5 | 83 |

**Table 3.** Classification results for the problem of predicting AF.

| **Optimization problem** | **Set of conditions** | **Number of conditions** | **Covering of negative objects** | **Covering of positive objects** | **Degree of conditions** | **Classification accuracy, %** |
|---|---|---|---|---|---|---|
| The objective function (1), delimitation (2) | neg. | 150 | 13 | 0 | 6 | 73 |
| | pos. | 150 | 0 | 13 | 5 | 58 |
| The objective function (1), delimitation (3) | neg. | 150 | 27 | 5 | 7 | 80 |
| | pos. | 150 | 5 | 27 | 6 | 68 |
| The objective function (1), delimitation (3) with building procedure | neg. | 150 | 27 | 4 | 9 | 80 |
| | pos. | 150 | 4 | 27 | 6 | 68 |
| The objective function (4), delimitation (3) | neg. | 82 | 16 | 5 | 6 | 67 |
| | pos. | 87 | 5 | 16 | 5 | 58 |

**Table 4.** Classification results for the problem of predicting PE.

| **Optimization problem** | **Set of conditions** | **Number of conditions** | **Covering of negative objects** | **Covering of positive objects** | **Degree of conditions** | **Classification accuracy, %** |
|---|---|---|---|---|---|---|
| The objective function (1), delimitation (2) | neg. | 152 | 18 | 0 | 7 | 62 |
| | pos. | 141 | 0 | 14 | 4 | 82 |
| The objective function (1), delimitation (3) | neg. | 152 | 36 | 5 | 7 | 68 |
| | pos. | 141 | 5 | 29 | 4 | 89 |
| The objective function (1), delimitation (3) with building procedure | neg. | 152 | 36 | 4 | 9 | 68 |
| | pos. | 141 | 4 | 29 | 5 | 89 |

| | | | | | | |
|---|---|---|---|---|---|---|
| The objective function (4), delimitation (3) | neg. | 67 | 16 | 5 | 6 | 62 |
| | pos. | 73 | 5 | 16 | 4 | 93 |

**Conclusion**

Tables 2-4 show that applying constraint (3) in the optimization model when searching for patterns allows finding patterns with higher coverage and a more accurate classifier is constructed on the basis of these patterns. Application of the considered approach is proper for solving problems with the existence of outliers and noises and with a lot of omissions in the data samples.

As a result of application procedure of growing patterns, the patterns with maximum covering and with higher degree are obtained, thus increasing reliability of the decisions made by the classifier constructed on the basis of these patterns. Increase in reliability of decisions occurs due to growth of informativeness of patterns. This is due to the fact that the number of the covered objects of own class remains unchanged but the number of the covered objects of another class decreases.

Application of the objective function (4) in the optimization model allows simplifying the classifier, thus reducing the number of patterns in it with respect to the complete set of patterns.

**References.**

[1] Hammer P L 1986 The Logic of Cause-effect Relationships *Lecture at the Int. Conf. on Multi-Attribute Decision via Operations Research-based Expert Systems (Passau, Germany)*

[2] Alexe S, Blackstone E. and Hammer P L 2003 Coronary Risk Prediction by Logical Analysis of Data *Annals of Operations Research* **119** 15-42

[3] Boros E, Hammer P L and Ibaraki T 2000 An Implementation of Logical Analysis of Data *IEEE Transactions on Knowledge and Data Engineering* 12(2) pp 292-306

[4] Hammer P L, Kogan A and Lejeune M 2004 Modeling Country Risk Ratings Using Partial Orders *RUTCOR Research Report* 24-2004

[5] Kuzmich R I and Masich I S 2012 Building a model for classification as decomposition of information patterns *Control Systems and Information Technology* **2(48)** 18-22

[6] Rastrigin L A and Freymanis E E 1988 Solving problems of different scaled optimization by methods of random search *Problems of random search* **11** 9-25

[7] Hammer P L and Bonates T 2005 Logical Analysis of Data: From Combinatorial Optimization to Medical Applications *RUTCOR Research Report* 10-2005

[8] Golovenkin S E, Gulakova T K, Kuzmich R I, Masich I S and Shulman V A 2010 Model of logical analysis for solving the problem of predicting myocardial infarction *Vestnik SibSAU* **4(30)** 68-73

[9] Kuzmich R I and Masich I S 2014 Modification of the objective function in the construction of patterns to increase the different conditions in the classification model *Control Systems and Information Technology* **2(56)** 14-18

[10] Golovenkin S E, Gorban A N, Shulman V A and Rossiev D A 1997 Complications of myocardial infarction: a database for testing recognition and prediction *Preprint* 1997.6 (Krasnoyarsk: Inst. of Computational Modeling SD RAS)

[11] Masich I S 2006 Approximate algorithms for finding the end-points for the problem of constrained pseudo-Boolean optimization *Vestnik SibSAU* **1(8)** 39-43

[12] Antamoshkin A N and Masich I S 2006 Heuristic search algorithms for monotone pseudo-boolean function conditional optimization *Engineering and automation problems* **1(5)** 55-61

[13] Antamoshkin A N and Masich I S 2007 Pseudo-Boolean optimization in case of unconnected feasible sets *Models and Algorithms for Global Optimization. Series: Springer Optimization and Its Applications* vol 4, ed A Törn and J Žilinskas (New York: Springer) pp 111-122

[14] Masich I S, Kuzmich R I and Kraeva E M 2011 *Logical analysis of data in classification problems* 2011612265 (Moscow: RosPatent)

IOP Conf. Series: Materials Science and Engineering **94** (2015) 012004 doi:10.1088/1757-899X/94/1/012004

# ANALYSIS OF FLOW ALONG SMOOTH SIDE OF DISC IMPELLER IN A PUMP

**E M Kraeva**

Department of Engines of Flight Vehicles, Siberian State Aerospace University named after academician M.F. Reshetnev,
31 "KrasnoyarskiyRabochiy" prospect, Krasnoyarsk, 660037, Russia.

E-mail: EM-Kraeva@yandex.ru

**Abstract.** We consider fluid flow in an axial gap around the smooth disc rotating in a closed cavity. According to the results of experiments on different working fluids in a wide range of operating modes for smooth discs, the formula is presented to determine the coefficient of fluid flow spin subject to a rate fluid flow from the periphery to the center. On the basis of the experimental data the empirical formula is obtained for the friction torque with respect to a rate fluid flow.

Modern pumps have very compact construction and low mass. They have an electric motor or a gas turbine as a rotor drive. Given the functional purpose, in addition to flowing parts, pumps contain many rotating cavities and channels with fluid or gas. These are varied hydraulic components: tip clearances and side cavities between disks of impellers and the body; axial gaps and cavities in contactless and hydrodynamic seals of the rotor; cavies between the rotor and the pump body [1, 2, 3, 4].

Achieving high performance and efficiency in a centrifugal pump, in addition to effective energy conversion in its flowing part, substantially depends on perfection of the hydraulic path with a complex internal interaction of fluid flows and its mutual influence on the flowing part of the centrifugal pump [1, 2, 4, 5, 6].

Let us consider the flow in the axial gap near a smooth disk rotating in a closed cavity. Imagine a smooth disc with a cylindrical surface of radius $R_H$, rotating in a closed chamber with an angular velocity $\omega$ in the absence of an expenditure fluids flow (figure 1). As a result of viscous forces impact, a circular movement arises in the chamber: near the disc a fluid moves from the center to the periphery, and near the wall it moves from the periphery to the center. Flow continuity is ensured by axial movement of the fluid: it moves from disk at the periphery in the boundary layer of thickness $\delta$, and it moves to the disc at the central part.

In systems of such type, the flow characteristics (angular velocity of the fluid and disk friction loss) significantly depend on a change of the Reynolds number $\mathrm{Re} = \omega R_H^2 / \nu$, relative axial gap $L/R_H$, and roughness.

Experimental researches of a fully wetted disc have shown that there are four flow regimes in the gap between the chamber wall and the rotating disk (figure 2):

- regime I - laminar fluid flow and merged boundary layers due to the small axial gap $L$. Circumferential velocity in the axial gap varies linearly;

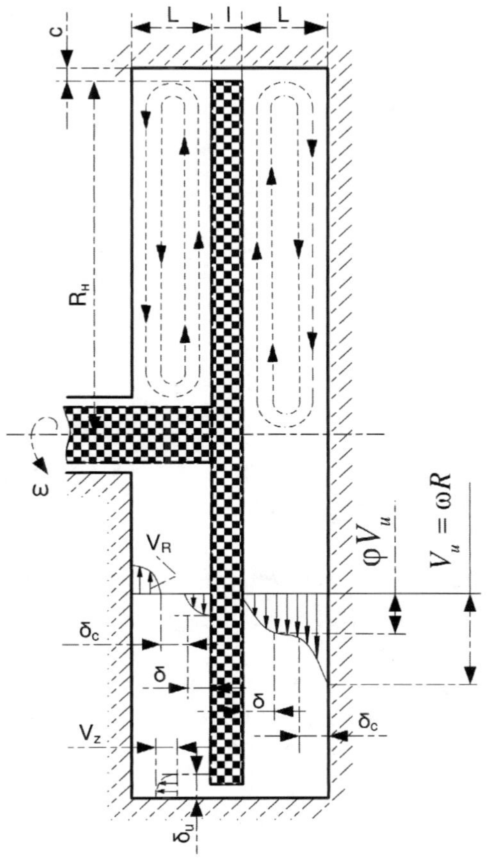


**Figure 1**. Disk rotating in a closed cavity.

- regime II – laminar fluid flow in the axial gap at separate boundary layers on the disc and the stationary wall of the chamber. Between the boundary layers, there is a flow core within which the circumferential velocity is constant;
- regime III – turbulent fluid flow with merged boundary layers due to the small axial gap $L$. Flow velocity of the fluid in the gap varies linearly;
- regime IV – turbulent fluid flow in the axial gap at separate boundary layers. Between the boundary layers, there is a flow core within which the circumferential component of the absolute velocity $V_u$ is constant.

Each flow regime characterized by its own relations that allow us to estimate the velocity of fluid flow in the gap and the moment resistance of the rotating disk.

For the Reynolds number Re > $3 \cdot 10^5$ the flow stops being laminar. And turbulent behavior of the flow is characterized at the pump elements in the real range of the viscosity of the working fluid for any value of the axial gap. Moreover, considering the large axial gaps in the high-speed centrifugal pump, i.e. $L \gg (\delta_c + \delta)$, we come to the conclusion that the analysis is required only for the regime IV of the fluids flow.

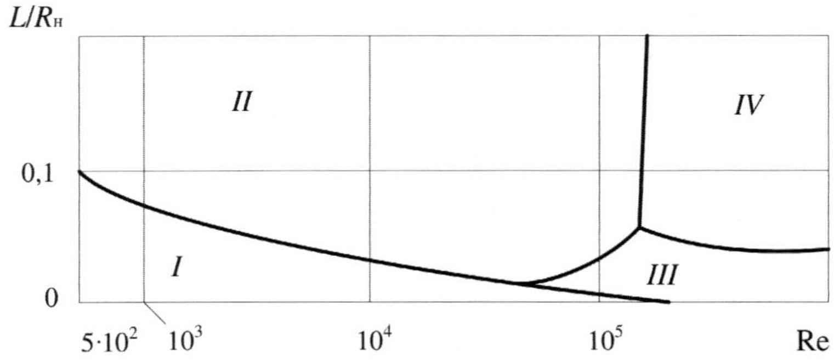


**Figure 2**. The area of existence of different flow regimes in the gap between the chamber wall and the rotating disc.

Such flow near a smooth disc has been studied in [7]. In this paper experimental data are presented, which show that the main part of fluid in the axial gap, called a flow core, rotates in the circumferential direction according to the law of solid body with constant angular velocity ω. The transition from the circumferential velocity of the fluid in the core to the speed at the disk and to the zero velocity at the wall occurs in a thin boundary layer between the core and the walls.

By the condition of equality of the moments on the wall and on the smooth disc, the value of the coefficient of the resistance moment for the smooth disk can be found by the formula [7]:

$$C_{м} = \frac{0.2335[(R_H + C)/R_H]^{23/5}\varphi^{9/5}}{4\mathrm{Re}^{1/5}}.$$

Where the coefficient φ is determined for small values $n = 100\frac{R_H + C}{R_H} - 1$ from the expression $1/\varphi = 2.06\,(1 + 0.014\,3n + 0.000\,141\,n^2 + \ldots)$.

For the axial gap limited by the ratio $L/R_H \leq 0.44$, what is typical for the construction of a centrifugal pump unit, with sufficient accuracy we can use the empirical formula for determining the coefficient of the resistance moment for the smooth side of the disk [7]:

$$C_{м} = \frac{0.151}{(L/R_H)^2\,\mathrm{Re}^{1.2}} + \frac{1.02 + L/R_H}{12(6 + L/R_H)\,\mathrm{Re}^{0.182}}. \quad (1)$$

In a pump unit the flow of fluid is forced and its flow rate is usually determined by the leak of a centrifugal pump or by conditions of heat removal from the seal cavities.

The investigation of the pressure distribution in the axial gap between the disk and the chamber wall in the presence of a radial flow was first held by A. A. Lomakin. The analysis of the results of measurement in the flow between the rotating disc and the chamber wall, carried out in [7], only generally confirmed the results of A. A. Lomakin.

The results of experiments on different working fluids [8] in a wide range of modes in operating of smooth discs allowed us to obtain the dependence to determine furling coefficient for the liquid flow at expendables fluid flows from the periphery to the center:

$$\varphi = 0.578 - 11V_R/(\omega R_H) - 0.0658\bar{L}, \qquad (2)$$

where $\bar{L} = L/R$.

For research of influence of expendables flows along the smooth side of the impeller disc on the pressure distribution over its radius, a series of experiments was held with smooth disks, which have the same geometric parameters with the tested impellers [9]. Experiments have shown that the experimental points obtained in testing a smooth disk with all the working fluids at the expendables flows in the axial gap from the center to the periphery for this range of parameters are well described by relation (2).

For the flow from the periphery to the center the expression is obtained

$$\varphi = 0.3865 + 95.3V_R/\omega R. \qquad (3)$$

For real values of clearance between the impeller and the chamber wall $L = 0.5\ldots4.0$ mm in the range of expendable parameter $V_R/\omega R = 0.005\ldots0.015$.

Under presence the flow rate, changes of the circumferential velocity in the gap influence on change in tension of friction and on the moment of resistance of the rotating disk [7]. This can be expressed by increasing the friction moment on the elementary circular area through width $\Delta R$:

$$\Delta M = 2\pi R^2 \Delta R \rho C_M (\omega - RV_u)^2/2. \qquad (4)$$

The comparison of the friction moments calculated by this equation for the conditions $V_u = \omega R/2$ (no flow rate) and $V_u \to 0$ (high flow rate) shows that for the second condition the friction moment on the disk is increased fourfold. On the basis of the experimental data the empirical formula was obtained for the friction moment taking into account an expendable flow:

$$M = \left( \frac{0.0187}{\sqrt[5]{\mathrm{Re}}} + \frac{0.08}{\sqrt{\omega R_H / V_{Rcc}}} \frac{L}{R_H} \right) \rho\, \omega^2 R_H^5 . \qquad (5)$$

The difference of the expressions (5) and (1) is that the second term in the bracket of equation (5) is the friction coefficient, due to the flow rate

$$C_{M\dot{V}} = \frac{0.08L}{R_H} \sqrt{\frac{V_{Rcc}}{\omega R_H}} . \qquad (6)$$

Accurate expression (6) for the expendable flow of water and air takes the form

$$C_{M\dot{V}} = 0.42 \cdot 10^3 \frac{(LR_H)^{0.75} Ga^{0.3}}{(\omega R_H / \nu_{Rcc})^{0,8}} , \qquad (7)$$

where *Ga* is the number of Galilee;

$$\omega R_H / V_{Rcc} = 2\pi R_H^2 L\omega \frac{1}{V}. \qquad (8)$$

In testing bladed disks of impellers for the modes, which are the same as the test modes for smooth disks, the value φ obtained that is larger than the value determined by the formula (2).

Let us represent the dependence $\varphi = f\,(V)$ calculated according to the formula (2) and the experimental points obtained in a test of a smooth disk (figure 3). It should be noted that a smooth disk is characterized by satisfactory convergence of the calculated dependences and the experimental points. However, for a pump impeller there is a significant difference between these data. This indicates that there is essential influence of vortex impulses on the hydrodynamics of the flow along the smooth side of the impeller. These impulses come from the blade side. At the same time the value φ for smooth side of the impeller depends on the flow rate through the cavity of the pump unit.

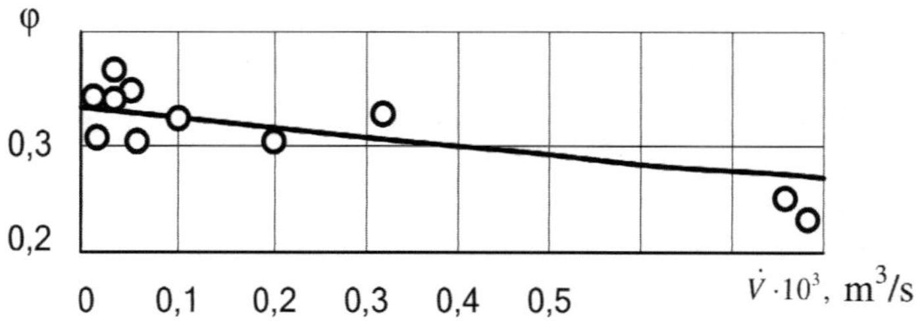


**Figure 3**. A graphical representation of φ on the flow rate in the axial gap for a smooth disk with $D_H$ = 80 mm [4].

For estimation of the vortex impulse exchange in the cavity of the pump unit, in the paper [6] a series of experiments is shown for smooth disks with different diameters (from 46 to 80 mm) with open impellers and hydrodynamic seals with a visor. Experiments were performed at the same axial and radial clearances with flow rate $\dot V = 0.05 \cdot 10^{-3}$ m³/s, which was caused by the necessity of heat transfer from the cavity of the pump unit.

Analysis of the obtained expressions and the calculation of φ for the different disks showed that these values for impellers with a visor and open impellers are different and at the same time similar quantities are more for smooth disks. Taking into account existing dependencies for smooth discs at zero and expendable flows, the value $\varphi_s$ for impellers can be determined by the expression

$$\varphi_s = k\varphi, \qquad (9)$$

where *k* is a coefficient determined empirically, it takes into account additional furling liquid in the smooth side of the impeller. As a result of processing of data obtained during the testing impellers and smooth disks with different sizes and different liquids at ω = 400 ... 2500 rad / s and flow rate $V = (0.05 \ldots 1) \cdot 10^{-3}$ $m^3$ / s, values $k = 2 \pm 0.1$ were obtained for an open impeller and values $k = 1.2 \pm 0.06$ were obtained for a hydrodynamic seal with visor.

Thereby, the results of researches conducted for the basic hydrodynamic parameters of the rotation cavities of pumps units allowed us to develop and experimentally prove the basic calculation expressions for calculating the flow along the smooth side of the disk impeller. The obtained analytical expressions describe the accumulated experimental data for the rational design of the flow part for cavities of rotation and can be used in the development of design solutions to provide the required level of energy excellence. All this contributes to efficiency and reliability of modern high-speed pumps units.

## References

[1] Kraeva E M and Kraev V M 2009 Hydrodynamic analysis and visualization of spatial flow in centrifugal force field *Preprint* 2009.2 (Krasnoyarsk: Inst. of Computational Modeling SD RAS) p 34
[2] Baybikov A S and Karahatyan V K 1982 *Hydrodynamics auxiliary paths bladed machines* (Moscow: Mashinostroyeniye) p 112
[3] Kraev M V, Ovchinnikov B V and Shapiro A S 1976 *Hydrodynamic seal high-speed shafts* (Moscow: Mashinostroyeniye) p 104
[4] Kraev M V and Taldykin Y A 1999 *Hydrodynamics cavities of rotation in rotating machines* (Krasnoyarsk: SAA Press) p 134
[5] Idelchik I E 1975 *Handbook of hydraulic resistance* (Moscow: Mashinostroyeniye) p 559
[6] Kraev M V 1983 *Theory and calculation of hydraulic pumps paths* (Krasnoyarsk: Krasnoyarsk Polytechnic Institute Press) p 100
[7] Dorfman L A 1960 *Pressure drop and heat transfer of rotating bodies* (Moscow: Fizmatgiz) p 260
[8] Kraeva E M 2009 On calculation of operating parameters of a low specific speed centrifugal pumps *Vestnik SibSAU* **2(23)** 168-170
[9] Kraeva E M 2011 *High-speed centrifugal pumps* (Krasnoyarsk: SibSAU Press) p 212

IOP Conf. Series: Materials Science and Engineering **94** (2015) 012005 doi:10.1088/1757-899X/94/1/012005

# AUTOMATED CONTROL AND DATA ACQUISITION SYSTEM FOR LITHIUM-ION ACCUMULATORS TEST BENCH*

**E A Mizrah, R V Balakirev, N V Shtabel**

Siberian State Aerospace University Named after Academician M.F. Reshetnev
31, Krasnoyarskiy Rabochiy Ave., Krasnoyarsk, 660037, Russia.

E-mail: shtabnik@gmail.com

**Abstract**. In the article authors describe architecture of an automated data acquisition system for lithium ion accumulators lifetime test bench. The system is based on high performance data acquisition modules and allows acquisition of data in the process of simultaneous testing of 12 accumulators, including current, voltage and temperature of each accumulator. LabView based software is used to control testing, registration of data and loading of test program into system modules.

## 1. Introduction

The lifetime testing of lithium-ion accumulators (LIA) is done during terrestrial accumulator development. In this way, it protects technology of LIA and its production. In the process of flight LIA production, electrical parameters of accumulator verified with acceptance testing. World market leaders use dynamic stress test (DST) to diagnostics of lifetime characteristics. This method based on sample accumulator selection from each flight production batch, and with high probability can measure lifetime characteristics and degradation of real accumulator. In addition to charge and discharge cycles test include temperature cycling which is done with heat chamber.

Accumulators test bench developed in SibSAU (Krasnoyarsk, Russia) [1, 2] is able to carry simultaneous testing of up to 12 LIA with continuous testing time up to 3 months.

The main tasks of data acquisition system are:

- test control in accordance with given program;
- measurement of accumulator parameters (current, voltage, temperature);
- heat chamber control;

* The research was done with the financial support of the Ministry of Education and Science of the Russian Federation in accordance with the agreement № 14.577.21.0082, unique identifier RFMEFI57714X0082

IOP Conf. Series: Materials Science and Engineering **94** (2015) 012005 doi:10.1088/1757-899X/94/1/012005

## 2. Test bench control system

Control system should provide continuous long-time testing of accumulators including current and voltage protections and emergency stop in case of energy breakdown.

Structure of control and data acquisition system is shown on figure 1. On figure 1 shown: TS – temperature sensors, CDM – charge-discharge modules, accumulators, National Instruments (NI) PXIe-8115 – control computer, NI PXI-6225, PXIe-4300 – data acquisition modules, NI PXI-8433/2 – RS-485 communication interface.

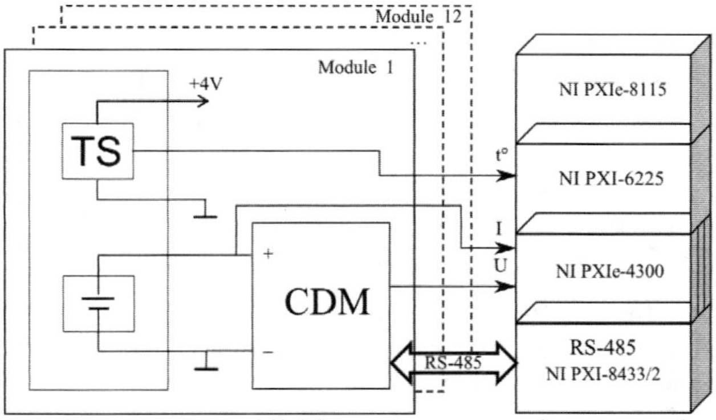


**Figure 1.** Structure of lithium-Ion accumulators test bench.

All CDM modules and control computer connected into one half-duplex RS-485 network. RS-485 protocol was selected because of its good noise protection and reliability. If necessary network can be expanded up to 32 devices (due to RS-485 protocol limitations) [3].

Main requirement for control system is continuous communication with each of 12 CDM's to obtain state information and provide fast response to accumulator state change. According to technical requirements data from all modules should be obtained each 250 ms.

Half-duplex network implies limitation of one master device in network which can send messages, all slave devices can only send response to master messages. In case of long-time testing it can be critical because during test there can be connection problems or computer software freezes. Even in this cases test should be completed or paused with respect to situation. Moreover, two CDM can work together with one accumulator, which means it should synchronize time of critical events such as charge to discharge and back transition.

Network with computer acting as master and CDM acting as slaves became fault-tolerant, because in case of computer problems system looses control.

To deal with this problems two methods was chosen: time-division multiplexing and autonomous CDM architecture.

Time-division multiplexing (TDM) is well-known technology of channel multiplexing where one communication channel can be divided into several subchannels with dedicated time slot (frame) [4]. Frames number can be fixed in case of static TDM or variable in case of dynamic TDM.

In case of computer communication with CDM to eliminate mandatory requirement of one master each device in network (every CDM and computer) has dedicated 16 ms time interval in which it allowed to send data. Each fixed time interval of 250 ms transmit cycle is repeated. One communication cycle time diagram is shown on figure 2.

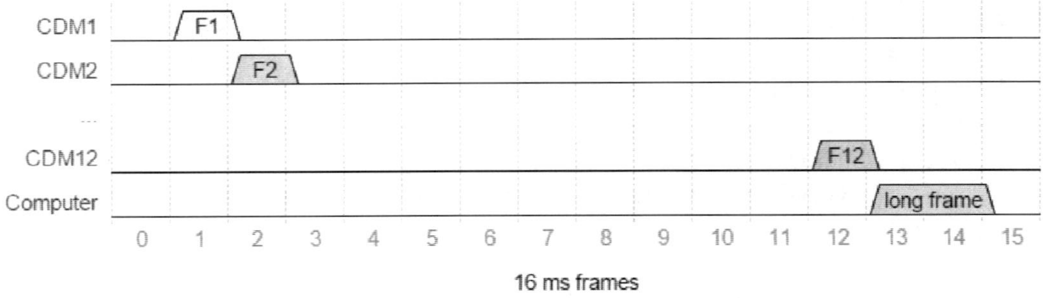


**Figure 2.** 250 ms communication cycle with TDM.

To synchronize devices with each other each device waits for packet start byte from device with previous address for time interval called "silence time" of 5 ms. If during this time it's not received, it means that device is not present in network and data can be sent.

This method allows to work without master device, but comes to new challenge: each device should react fast enough to previous device packet not to lose it's time frame. It's easy to obtain in CDM control devices which are based on microcontrollers and allow to get sub-millisecond reaction, but in case of computer it's not true.

In our case computer works on non-real-time operating system, which means that communication task can't guarantee response time [5]. In case of Windows operating system it means that even if task is working with highest priority and it send data to communication port driver in required time, it can't guarantee that driver will process data quick enough [6].

During test it was shown that time needed to process data from driver buffer to port output can easily grow with increase processor load even in case of highest task priority. In test with 10% processor load maximum response time was 30 ms which is sufficient for wider time frame of 58 ms dedicated for the computer. But in case of 80% and more processor load maximum response time grown to more than 100 ms. It is unacceptable because it leads to transmit overlap over devices resulting in communication error. Example of delayed frame shown on figure 3.

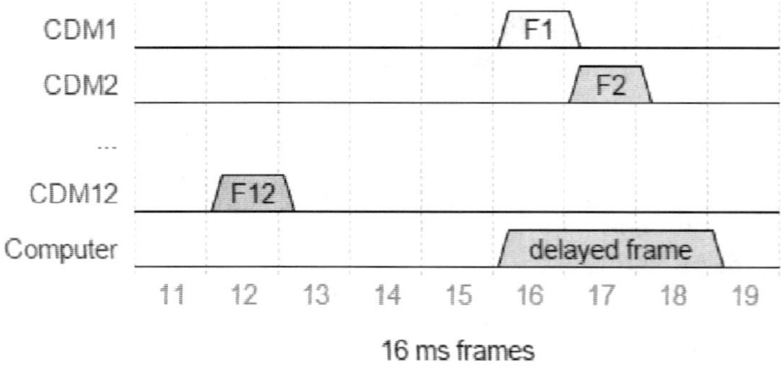


**Figure 3.** Communication cycle with delayed computer frame.

Problem of real-time response has several solutions, one of them is using real-time operating system on computer, but in National Instruments configuration it requires one more additional computer – one for communication task with real-time operating system and one for user interface and control with general purpose operating system. System became complicated in this case, communication and user interface layers should communicate between each other and has to be fully separated modules.

More straightforward solution is to use dedicated microcontroller pairing device (MPD) between computer and CDM to buffer packets and transmit them only in allowed time frames. It means that for computer system remains transparent, it can send packets at any time, because packets will remain in MPD buffer before next allowed data frame. Responses to computer transmitted directly without delay, because in given solution communication interface between computer and MPD is full-duplex with separated receive and transmit lines. Example of data communication frame with MPD shown on figure 4.

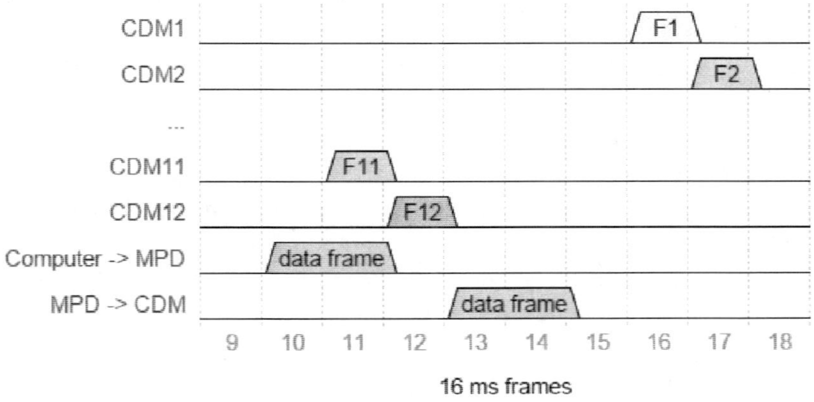


**Figure 4.** Communication cycle with microcontroller pairing device.

During test operator inputs test program via user interface, which contains sequence of LIA charge-discharge cycles and test steps. Each step includes information about required LIA voltage, current, charge or discharge method (with constant current, changing current, constant power), duration and repetition count of each step. Test program read from text file, written in special algorithmic language, which is converted to machine form to load program into CDM. Currently next features are supported:

- Charge and discharge with constant or linear changing current
- Loops (up to 65535 repetitions)
- Simple conditions based on loop counter
- Dynamically changing current and voltage protection
- Delay
- Special commands to measure accumulator capacitance

Text to binary format converter and communication protocol with CDM done in native LabView language.

Autonomous CDM architecture means that full test program including all steps stored inside CDM, and even in case of communication failure with computer it will continue to work and will be able to correctly shut down.

## 3. Test bench data acquisition system

Data acquisition system consist of two types of modules:

- NI PXIe-4300 modules with 8 isolated analogue input channels
- NI PXI-6225 module with 80 non-isolated analogue channels

Isolated analogue input channels used to measure current and voltage of accumulator with high accuracy. Non-isolated analogue input channels used to measure accumulator temperature.

One of the main requirements to test bench is accuracy, which can be obtained by using precise sensors and measurement devices. Accumulator voltage range is -5 to 5 V and can be measured directly by PXIe-4300. Voltage measurement error should be lower than 5 mV. Current measurement is more complicated because of high current range up to 160 Amperes and accuracy requirements:

- 30 mA in range 0-5 A
- 50 mA in range 5-50 A
- 200 mA in range 50-160 A

To match this requirements LEM Ultrastab IT200-S current sensor was chosen, it has measurement accuracy of 0.1% [7].

Filtering of current and voltage signals is necessary because of high-frequency noise coming from two convertors inside CDM which are working on frequencies of 100 kHz and 10 kHz. Two filters are used for filtering: 10 kHz 2-nd order digital Butterworth low-pass filter and additional 1 kHz analogue first-order Butterworth low-pass filter. It allows to efficiently reduce noise on analogue inputs. In addition to filtering averaging is used, with 100 averages on 1 kHz sample rate [8]. Results of current measurement analysis for 14 modules shown on figure 5. Voltage measurement analysis for 14 modules shown on figure 6.

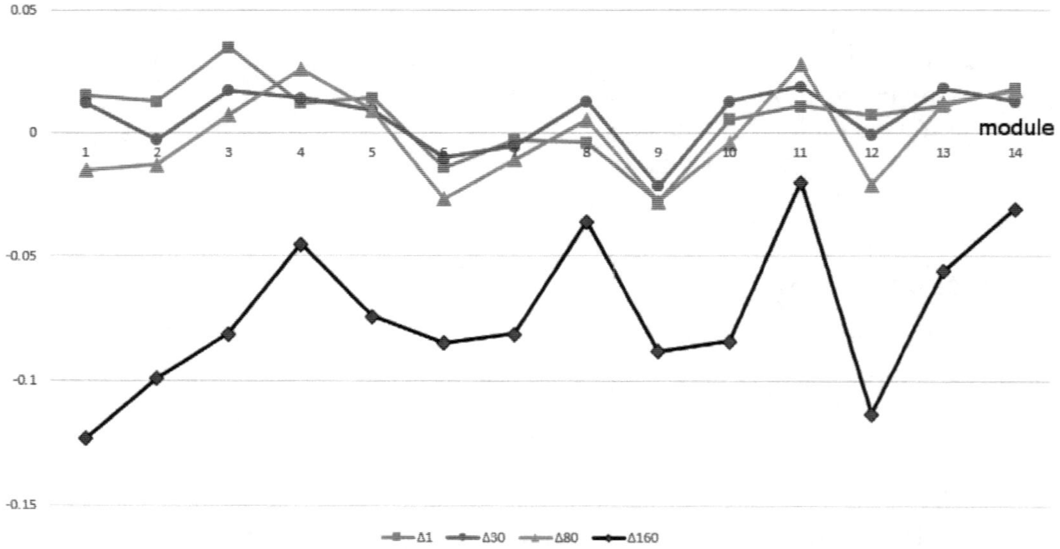


**Figure 5.** Current measurement error for currents of 1A, 30A, 80A, 160A.

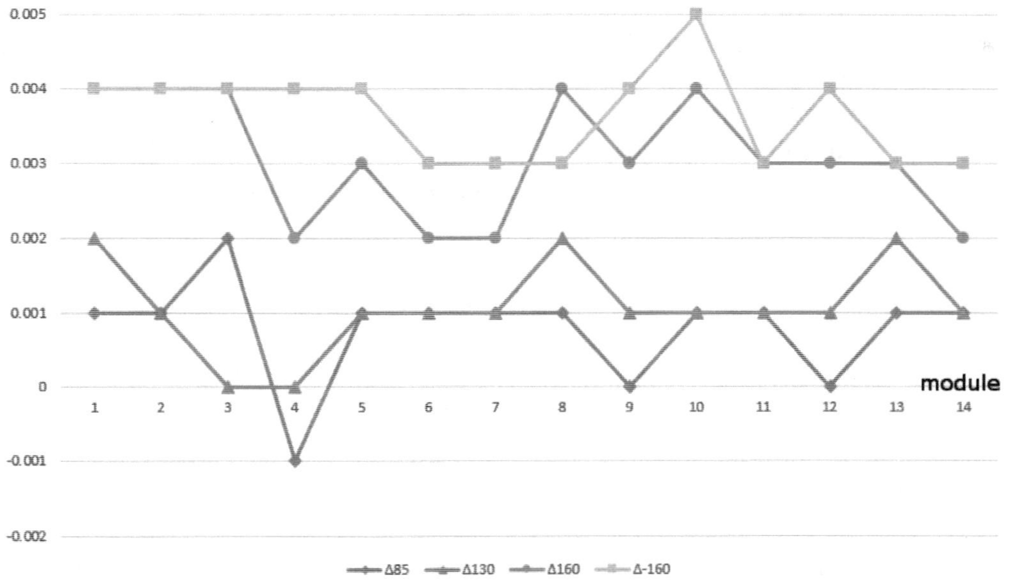


**Figure 6.** Voltage measurement error for currents of 85A, 130A, 160A.

From measurements on figure 5 and figure 6 we can see that measurement error in all ranges is lower than allowable.

To store data TDMS file format was chosen [9], because of its high portability, ability to store metadata together with values, ready program modules in LabView to directly stream collected data to file.

Programmatically data acquisition done in two loops, one for data collection into buffer and one for data writes to disk, it allow continuous data collection every 100 ms without interruption and eliminates dependence on disk drive performance.

## 4. Conclusion

Presented architecture of modular accumulator test bench allows to build scalable system for near any type of known accumulators besides lithium-ion. Time division multiplexing used in communication system is expandable and can be used for more than 20 modules depending on communication baud rate. Data acquisition system can be also extended to match requirements. Accumulators test bench presented in the article supposed to use in accumulators production quality control.

## References

[1] Mizrah E A, Kopylov E A and Balakirev R V 2012 *Accumulator cell testing device* Russia Utility model patent 123530

[2] Volochaev M N and Mizrah E A 2013 Charge characteristics simulator for spacecraft power system testing *Proceed. of XVII International Science conference in memory of M.F.Reshetnev* (Krasnoyarsk: SibSAU) **1** 196

[3] Axelson J 1998 *Serial port complete: programming and circuits for RS-232 and RS-485 links and networks* (Madison, WI: Lakeview Research)

[4] Hanrahan H E 2005 *Integrated Digital Communications* (Johannesburg: University of the Witwatersrand).

[5] Tanenbaum A S 2008 *Modern operating systems* (Upper Saddle River, N.J.: Pearson

Prentice Hall)
[6] Orwick P and Smith G 2007 Developing drivers with the Windows driver foundation (Redmond, Wash.: Microsoft Press)
[7] 2014 *LEM Ultrastab IT-200 datasheet* (http://www.lem.com/docs/products/it_200-s_ultrastab.pdf)
[8] Williams A B and Taylor F J 2006 *Electronic filter design handbook* (New York: McGraw-Hill)
[9] 2015 *The NI TDMS File Format* (http://www.ni.com/white-paper/3727/en/)

IOP Conf. Series: Materials Science and Engineering **94** (2015) 012006 doi:10.1088/1757-899X/94/1/012006

# RESEARCH OF CONTROL SYSTEM STABILITY IN SOLAR ARRAY SIMULATOR WITH CONTINUOUS POWER AMPLIFIER OF PARALLEL TYPE*

**E A Mizrah, S B Tkachev, N V Shtabel**

Siberian State Aerospace University Named after Academician M.F.Reshetnev
31, Krasnoyarskiy rabochiy Ave., Krasnoyarsk, 660037, Russia.

E-mail: shtabnik@gmail.com

**Abstract**. Solar array simulators are nonlinear control systems designed to reproduce static and dynamic characteristics of solar array. Solar array characteristics depend on illumination, temperature, space environment and other causes. During on-earth testing of spacecraft power systems there is a problem reaching stable work of simulator with different impedance loads in wide range load regulation. In the article authors propose a research method for absolute process stability in solar array simulators and present results of absolute stability research for solar array simulator with continuous parallel type power amplifier.

## 1. Introduction

Solar array simulators (SAS) [1-8] are widely used during on-earth testing of space power systems. Simulators should reproduce nonlinear current-voltage characteristics [9] (I-V curve) and impedance (admittance) of solar array [10] with given accuracy. During test simulator should reproduce I-V curve characteristics: short circuit current, open circuit voltage, I-V curve shape (figure 1) which can vary in wide range.

* The research was done with the financial support of the Ministry of Education and Science of the Russian Federation in accordance with the agreement № 14.577.21.0082, unique identifier RFMEFI57714X0082

IOP Conf. Series: Materials Science and Engineering **94** (2015) 012006 doi:10.1088/1757-899X/94/1/012006

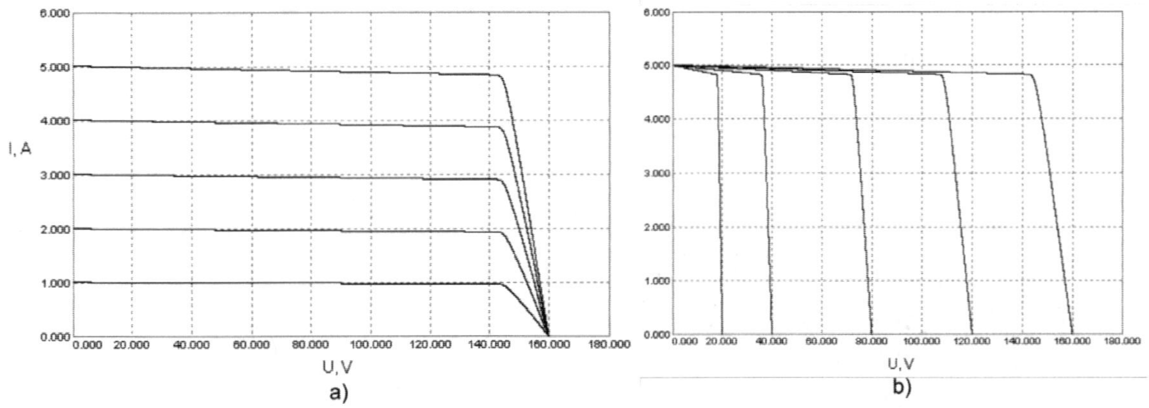


**Figure 1.** SAS I-V curve with voltage (a) and current (b) regulation.

To reproduce solar array admittance-frequency characteristic in range from 0 to 1000 kHz SAS should be implemented with continuous control. In this case SAS can be built using two methods:

- obtaining I-V curve in control circuit
- obtaining I-V curve in load circuit.

Advantages and disadvantages of each method of I-V curve obtaining method reviewed by authors in article [11]. Simulators with control circuit I-V curve obtaining method usually based on control system with nonlinear functional feedback and continuous power amplifier (CPA).

To conform given requirements on solar array static and dynamic characteristics reproduction accuracy, problem of SAS control system absolute stability in condition of wide range nonlinear I-V curve regulation should be solved. SAS with CPA connected in series to load [12] are widely used. SAS with CPA of parallel type are less studied [13].

The purpose of the article is to give survey on absolute stability of electrical process in SAS with parallel type CPA.

## 2. Preliminary

In accordance with structure (figure 2), SAS is a load current stabilizer with nonlinear I-V curve forming feedback based on functional transformer (FT). I-V curve regulation done by using current scale device (CSD) and voltage scale device (VSD). SAS open circuit voltage range is 20-160 V, short circuit current range is 1-5 A.

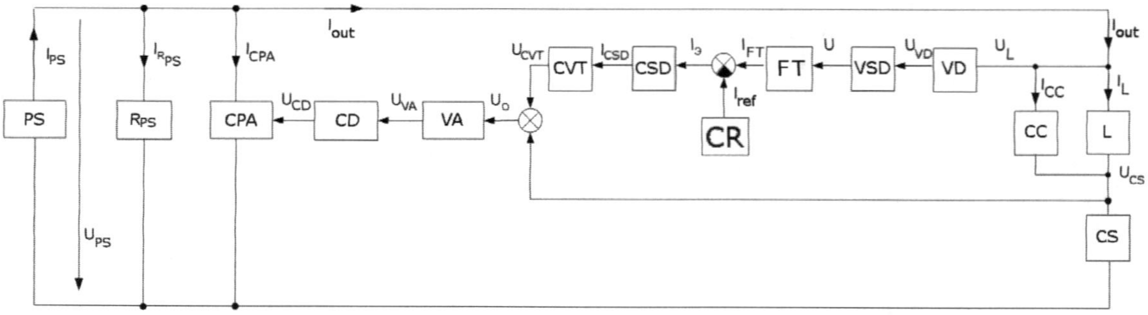


**Figure 2.** Structure of solar array simulator with parallel CPA.

IOP Conf. Series: Materials Science and Engineering **94** (2015) 012006 doi:10.1088/1757-899X/94/1/012006

On figure 2 shown next devices: PS – power source (current source), RPS – PS internal resistance, CPA – continuous power amplifier, CD- serial correction device, VA – error voltage amplifier, CVT – current to voltage transformer, CSD – current scale device, CR – current reference, FT – functional transformer, VSD – voltage scale device, VD – voltage divider, CC – correction circuit, L – load, CS – current sensor.

It is advisable to use absolute stability criteria by Naumov and Cypkin [14] to analyse absolute stability of processes in nonlinear control systems. Due to criteria, functional scheme of SAS should be represented (figure 3) to contain linear part with transfer function (TF) $W_{lin}(s)$ and nonlinear functional feedback $F_{ft}(U)$.

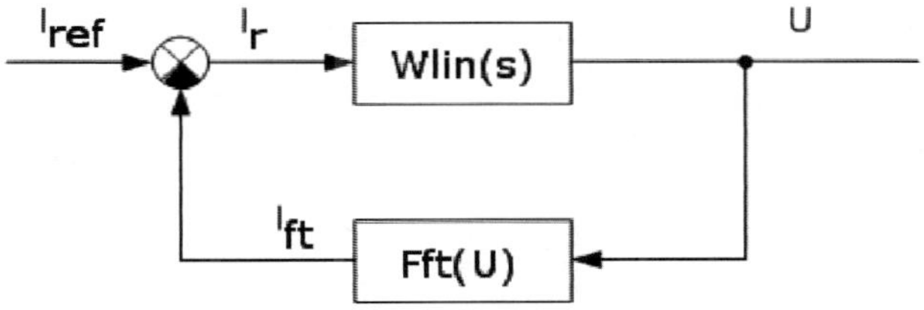


**Figure 3.** Generalized functional scheme of SAS with CPA of parallel type.

According to figure 2 for each SAS device, we can obtain symbolic equation, describing processes in the device.

Let's represent PS in form of current source $I_{PS}(s)$ with internal resistance $Y_{PS}(s)$:

$$Y_{PS}(s) = \frac{I_{RPS}(s)}{U_{PS}(s)} = \frac{1}{R_{PS}(s)}$$

Then we can write:

$$I_{PS}(s) = I_{RPS}(s) + I_{CPA}(s) + I_{OUT}(s)$$

$$I_{OUT}(s) = I_{CC}(s) + I_L(s)$$

According to equations, we can compose functional scheme of simulator (figure 4).

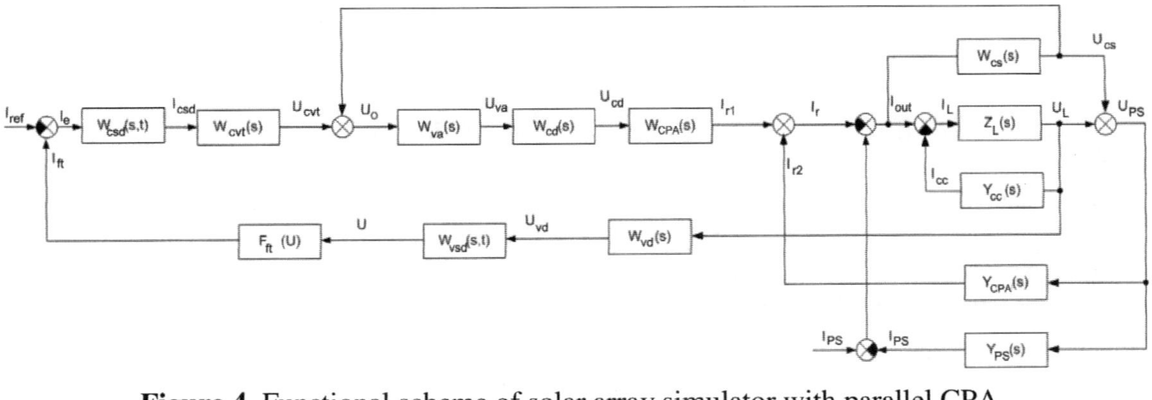


**Figure 4.** Functional scheme of solar array simulator with parallel CPA.

## 3. Methodics

Methodic of correction device and correction circuit synthesis providing required admittance-frequency characteristic and stability of linearized SAS control system given in [15].

According to Naumov-Cypkin stability criteria the authors proposed next methodic of SAS stability research:

- Functional scheme should be represented in one loop form (figure 3) with separation of linear part $W_{lin}(s)$ and nonlinear unit $F_{ft}(U)$.
- Determine devices transfer functions using mathematical description, numeric experiments in circuit simulator (for example, Micro-CAP), or using empirical methods. Determine maximum values of scaling devices gain.
- Determine worst-case maximum gradient value $k_{FTmax}$ for nonlinear unit (functional converter). Due to functional scheme:

$$k_{FT\max} = \left(\frac{\partial I_{FT}}{\partial U}\right)_{\max} = \frac{k_{IVC\max} W_{CS}(0)}{W_{VSD\min}(0) W_{CSD\min}(0) W_{CVT}(0) W_{VD}(0)} \tag{1}$$

Where maximum slope of I-V curve defined by:

$$k_{IVC\max} = \left(\frac{\partial I}{\partial U}\right)_{\max}$$

$W(0)$ is a transfer coefficient of the element at frequency $\omega=0$.

- Draw normalized Bode magnitude plot (BMP)

$$L_0(\omega) = 20\lg\left(k_{FT\max}\left|W_{lin}(j\omega)\right|\right)$$

and Bode phase plot (BPP) $\varphi_{lin}(\omega)$ of linear part.

- Draw BMP of critical transfer coefficient

$$L_{CR}(\omega) = -20\lg\left(\left|\cos(\phi_{lin}(\omega))\right|\right)$$

- Verify criteria of absolute stability:

$$L_{CR}(\omega) > L_0(\omega) \tag{2}$$

in frequency range where next requirement is met:

$$-(4m+1)\frac{3\pi}{2} \geq \phi_{lin}(\omega) \geq -(4m+1)\frac{\pi}{2} \quad m = 0,1,2,... \tag{3}$$

If normalized BMP $L_0(\omega)$ will not cross BMP of critical coefficient $L_{CR}(\omega)$, processes in SAS will be asymptotically stable in case of I-V curve deformation within given current and voltage ranges.

To get transfer function of linear part Mason formula used [16]. Analyzing part of scheme (figure 4) which does not contain nonlinear unit $F_{ft}(U)$ with input signal $I_E(s)$ and output signal $U(s)$ on FC input, we can write:

$$W_{lin}(s) = \frac{I_E(s)}{U(s)} = \frac{H_1(s)}{\Delta(s)} \tag{4}$$

Where

$$H_1(s) = W_{CSD\max}(s) W_{CVT}(s) W_{CP}(s) Z_L(s) W_{VD}(s) W_{VSD\max}(s)$$

$$W_{CP}(s) = W_{VA}(s) W_{CD}(s) W_{CPA}(s)$$

$$\Delta(s) = 1 - \left(H_{11}(s) + H_{12}(s) + H_{13}(s) + H_{14}(s) + H_{15}(s) + H_{16}(s)\right) + \left(H_{11}(s) + H_{12}(s) + H_{14}(s)\right)$$

$$H_{11}(s) = -W_{NL}(s) W_{CS}(s)$$

$$H_{12}(s) = -W_{CS}(s) Y_{CPA}(s)$$

$$H_{13}(s) = -Z_L(s) Y_{CPA}(s)$$

$$H_{14}(s) = -W_{CS}(s) Y_{PS}(s)$$

$$H_{15}(s) = -Z_L(s) Y_{PS}(s)$$

$$H_{16}(s) = -Z_L(s) Y_{CC}(s)$$

To determine transfer function for devices without analytical description authors created SAS circuit model in Micro-CAP circuit simulator (figure 5). The circuit model is equivalent to SAS structure shown on figure 2. In the circuit model next PSPICE models are used: AD8066 operational amplifier (X1-X3), complementary pair of IRF7309 FET (VT1-VT2), IRF740 (VT3), bipolar transistors BC107 (VT7-VT9), BC309 (VT4-VT6).

Using frequency analysis program (AC-Analysis) of Micro-CAP together with Invfreqs MATLAB function we can determine transfer functions:

$$W_{CSD\max}(s) = 1$$

$$W_{CVT}(s) = \frac{4000}{\left(9.6 \cdot 10^{-9} \cdot s + 1\right)}$$

$$W_{CP}(s) = \frac{216}{\left(1.6 \cdot 10^{-5} \cdot s + 1\right)\left(1.6 \cdot 10^{-8} \cdot s + 1\right)\left(1.3 \cdot 10^{-8} \cdot s + 1\right)}$$

$$W_{VD}(s) = 0.05$$

$$W_{VSD\max}(s) = \frac{8}{\left(2.65 \cdot 10^{-8} \cdot s + 1\right)}$$

$$W_{CS}(s) = 0.4$$

$$Y_{CPA}(s) = \frac{303 \cdot 10^{-6}(2.87 \cdot 10^{-4} \cdot s + 1)}{\left(3.18 \cdot 10^{-8} \cdot s + 1\right)\left(3.18 \cdot 10^{-8} \cdot s + 1\right)}$$

$$Y_{CC}(s) = \frac{C_{CC} s}{(C_{CC} R_{CC} \cdot s + 1)} = \frac{2 \cdot 10^{-6} s}{2 \cdot 10^{-6} \cdot 0.1 s + 1}$$

$$Y_{PS}(s) = \frac{\left(C_{PS} R_{PS} \cdot s + 1\right)}{R_{PS}} = \frac{5 \cdot 10^{-3} s + 1}{10 \cdot 10^{3}}$$

IOP Conf. Series: Materials Science and Engineering **94** (2015) 012006 doi:10.1088/1757-899X/94/1/012006

Current scale device has maximum transfer coefficient $k_{CSD\max} = 1$ with maximal current $I_{SC} = 5$ A. Voltage scale device has maximum transfer coefficient $k_{VSD\max} = 8$ with minimal voltage $U_{OC} = 20$ V.

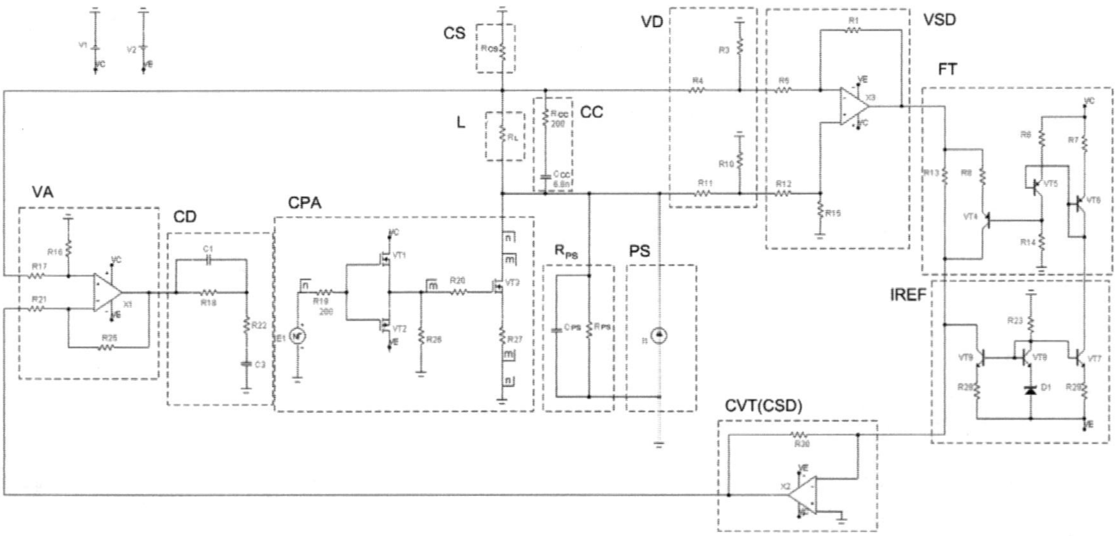


**Figure 5.** Micro-CAP circuit model of SAS with parallel CPA.

Maximum I-V curve gradient is

$$k_{IVC\max} = 0.35 \text{ A/V}$$

By substitute of static TF coefficient in equation (1) we can get maximum functional transformer gradient

$$k_{FT\max} = 700 \cdot 10^{-6} \ \mu\text{A/V}$$

One of the main requirements to SAS is stable work with PWM based spacecraft electrical power system converters, namely with step-down or step-up converters. Due to step-down input capacitance filter, input impedance of converter has generally capacitive form. Input impedance of step-up converter has inductance form [17]. Also during SAS verification resistance load used to obtain I-V curve. So, to verify absolute stability criteria we should review three types of load:

- Resistant

$$Z_R(s) = R_L$$

- Resistant-capacitive

$$Z_C(s) = \frac{R_L}{R_L C_L \cdot s + 1}$$

- Resistant-inductive

$$Z_L(s) = L_L \cdot s + R_L$$

## 4. Results

Now it is possible to do a verification of absolute stability criteria. In MathCAD we draw normalized BMP $L_0$ and BMP of critical coefficient $L_{CR}$, and BPP of linear part for each of three types of load.

Analysis of absolute stability criteria fulfillment in case of resistant load done with next assumptions. Maximal transfer coefficient of nonlinear functional converter unit reached in open circuit mode or maximal load resistance, which determined by feedback voltage divider resistance connected parallel to infinite load resistance. From that

$$R_L = R_{VD} = 4 \text{ kOhm}$$

To compare fulfillment of absolute stability criteria in case of different load types, we analyze SAS performance near I-V curve maximum power point, because voltage converters usually works in maximum power point (capacitive-resistant and inductive resistant loads). Maximum power point obtained with load resistance of $R_L = 28.5$ Ohm.

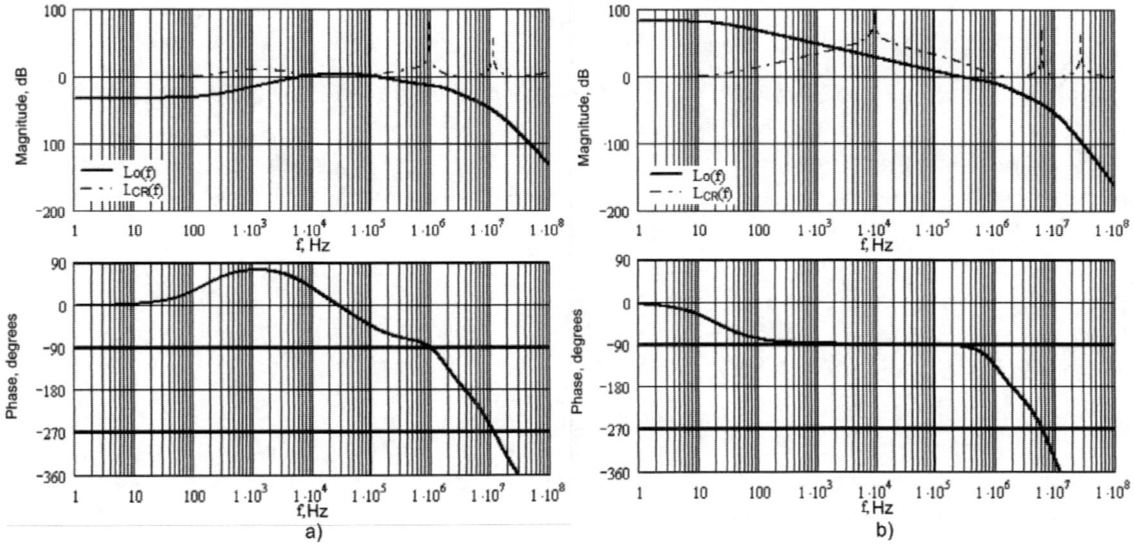


**Figure 6.** Bode plot of SAS with parallel CPA for resistant load $R_L$=28,5 Ohm (a), $R_L$=4 kOhm (b).

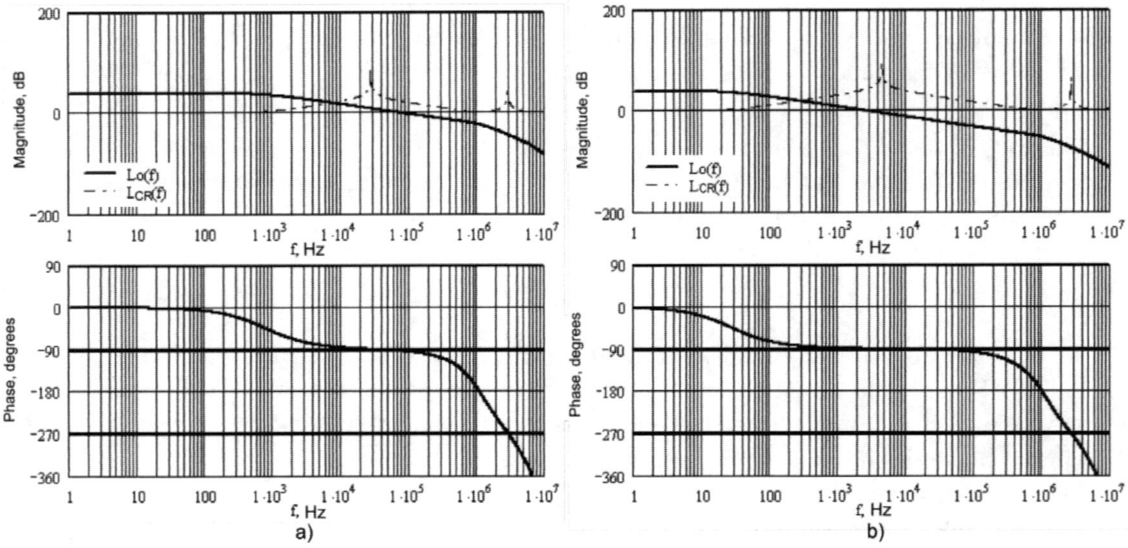


**Figure 7.** Bode plot of SAS with parallel CPA for capacitive load $C_L$=5 μF (a), $C_L$=150 μF (b).

Bode plot analysis (figure 6) shows that requirements of absolute stability criteria (2) and (3) fulfilled, meaning that processes in SAS in case of resistant load are asymptotically stable during I-V curve regulation within given ranges.

Verification of absolute stability criteria fulfillment in case of capacitive-resistant load (figure 7) done for two $C_L$ values. Load capacitance $C_L$ = 5 μF, $C_L$ = 150 μF and $R_L$ = 28.5 Ohm.

Label with $\Delta L$ stability reserve:

$$\Delta L = (L_{CR}(\omega) \ - \ L_0(\omega))_{min}$$

From figure 7 we can see that in case of $C_L$ = 150 μF, normalized BMP $L_0(\omega)$ passes much below BMP $L_{CR}(\omega)$ of critical coefficient, than in case of $C_L$ = 5 μF, i.e. requirement $\Delta L_5 < \Delta L_{150}$ fulfilled.

From above we can conclude that absolute stability criteria (2) and (3) fulfilled in case of capacitive-resistant load with different capacitance and increase of load capacitance comes to increase of stability reserve $\Delta L$.

We can verify absolute stability criteria in case of inductive-resistant load (figure 8) with inductance $L_L$=30 μH, $L_L$=200 μH and $R_L$=28.5 Ohm. Value of inductive part is the same order with input choke of step-up stabilizer of converter.

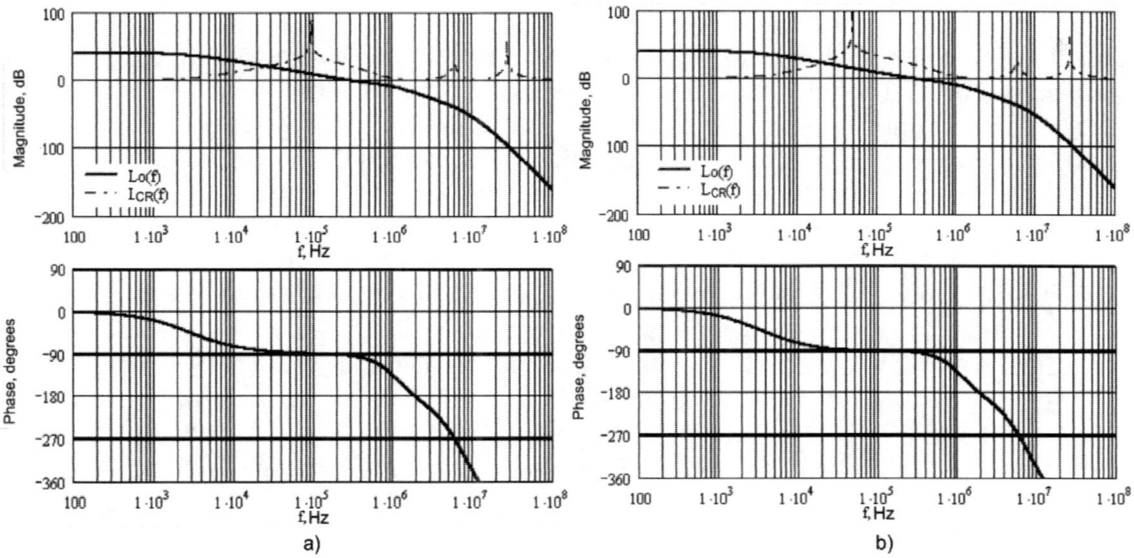


**Figure 8.** Bode plot of parallel type CPA SAS for inductive load $L_L$=30 μH (a), $L_L$=200 μH (b).

From analysis of Bode plot (figure 8) we can conclude that with increase of inductance value processes in SAS became absolute stable furthermore, increase of load inductance comes to frequency range widening as of expression (3), i.e. it has effect on linear part BPP and critical coefficient $L_{CR}$, but stability reserve ΔL remains the same.

On figure 9 presented transient process of output current and voltage for one channel of solar array simulator in wide-range load switching mode on current segment of I-V curve from short circuit to maximum power point of I-V curve with $U_{OC}$=160V and $I_{SC}$=5A. From figure 9 we can see that processes are stable. Time of transient process of 60 microseconds is equal to transient time of real solar array.

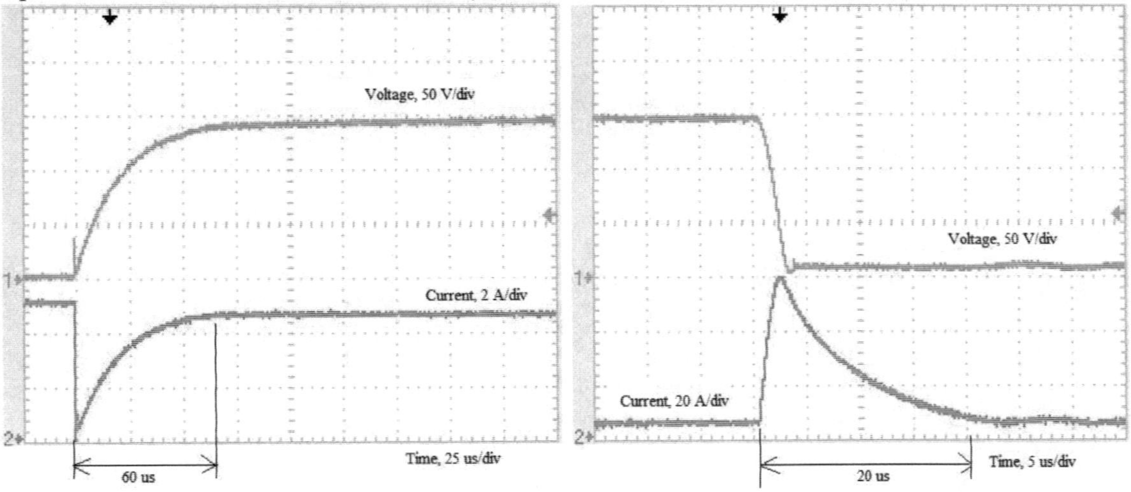


**Figure 9.** Transient process of SAS with parallel CPA in wide-range load switching mode on current segment of I-V curve from short circuit to maximum power point.

IOP Conf. Series: Materials Science and Engineering **94** (2015) 012006 doi:10.1088/1757-899X/94/1/012006

## 5. Conclusion

Summarizing above analysis we can obtain next conclusion about absolute stability of solar array simulator with parallel CPA: processes in solar array simulator can be considered stable in case of I-V curve given range regulation with three typical load types if simulator reproducing admittance-frequency characteristic of solar array with given accuracy.

## References.

[1] Robson R R 1975 7.5 kW Solar array simulator *Proc. American Institute of Aeronautics and Astronautics and Society of Automotive Engineers 11th Propulsion Conference* (Anaheim, Calif., Sept 29-Oct 1 1975) pp 7–10

[2] Haeberlin H, Borgna L, Gfeller D, Schaerf P and Zwahlen U 2008 Development of a fully automated PV array simulator of 100 kW *Proc. 23rd European Photovoltaic Solar Energy Conference* (Valencia)

[3] Agilent E4350B, E4351B Modular Solar Array Simulators datasheet www.agilent.com

[4] Elgar Solar Array Simulators. www.elgar.com

[5] Mishin V N, Pchelnikov V A, Bubnov O V and Kremzukov J A 2014 Electrical solar array simulator Utility model patent 144248 Russian Federation (Tomsk, Tomsk state university).

[6] Gonzalez S, Kuszmaul S, Deuel D and Lucca R 2010.PV Array Simulator Development and Validation *Proc. 35th IEEE Photovoltaic Specialists Conference* (Honolulu, Hawaii, USA 20 - 25 June 2010) pp 2849–52.

[7] Sidorov A S and Balakirev R V 2007 Two-channel solar array simulator with automated control *Proc. XV Tupolev readings: International conference (Kazan)* **4** 95–96

[8] Mizrah E A, Sidorov A S, Evenov G D and Shanavrin V S 2008 Solar array simulation device Patent 2342815 Russian Federation.

[9] McEvoy A J, Markvart T and Castaner L 2012 *Solar Cells Materials, Manufacture and Operation 2nd ed.* (Oxford: Academic) pp. 501–29.

[10] Illanes R, De Francisco A, Núñez F, De Blas M, García A and Torres J L 2014 Dynamic simulation and modelling of stand-alone PV systems by using state equations and numerical integration methods *Applied Energy* **135** 440–49

[11] Mizrah E A, Sidorov A S 2007 Modeling of solar array simulator with cascade connection of continuous and impulse power amplifier *Vestnik SibGAU* (Krasnoyarsk: SibSAU) **14** pp 7–12

[12] Mizrah E A and Sidorov A S 2003 Research of solar array simulator with continuous parallel power amplifier *Vestnik SibGAU* (Krasnoyarsk: SibSAU) **47** pp 12–15

[13] Naumov B N 1972 *Theory of nonlinear automatic systems. Frequency methods* (Moscow) pp 89

[14] Mizrah E A, Sidorov A S and Balakirev R V 2006 Synthesis of required admittance-frequency characteristics in solar array simulator *Vestnik SibGAU* (Krasnoyarsk: SibSAU) **9** pp 55-57

[15] Mason S and Zimmerman H 1963 *Electronic circuits, signals, and systems* (N.Y.: John Wiley and Sons)

[16] Patel M R 2005 *Spacecraft Power System* (N.Y.: CRC Press)

[17] Erickson R W 1997 *Fundamentals of Power Electronics* (N.Y.: Chapman and Hall)

IOP Conf. Series: Materials Science and Engineering **94** (2015) 012007 doi:10.1088/1757-899X/94/1/012007

# THE BIOMETRIC-BASED MODULE OF SMART GRID SYSTEM*

**E Engel[1], I V Kovalev[2], A Ermoshkina[3]**

[1] Associate Professor, First Russian doctoral degree in computer sciences, Katanov State University of Khakassia, Abakan, RF,

[2] Professor, Second Russian doctoral degree in computer sciences, Siberian State Aerospace University, Krasnoyarsk, RF,

[3] Undergraduate student, Katanov State University of Khakassia, Abakan, RF

E-mail: Angel@khsu.ru

**Abstract.** Within Smart Grid concept the flexible biometric-based module base on Principal Component Analysis (PCA) and selective Neural Network is developed. The formation of the selective Neural Network the biometric-based module uses the method which includes three main stages: preliminary processing of the image, face localization and face recognition. Experiments on the Yale face database show that (i) selective Neural Network exhibits promising classification capability for face detection, recognition problems; and (ii) the proposed biometric-based module achieves near real-time face detection, recognition speed and the competitive performance, as compared to some existing subspaces-based methods.

## 1. Introduction

The Smart Grid vision embraces the idea of a society of connected smart devices and systems that self-assemble and seek operational efficiencies and reliable performance [1]. Smart grid will aid in the integration "micro-generation" facilities into the intelligent system. How successful the Smart Grid project can be largely depends on how well it defends against remote network-based attacks. User authentication for accessing the Smart Grid is the first and strongest line of defense against these types of attacks. Modern password based authentication mechanism has been proven inadequate. It is believed that biometric authentication will significantly improve the security of the Smart Grid network. In this paper we propose face recognition technology base on selective Neural Network to authenticate users accessing the Smart Grid. Considering Smart Grid technology we are focused on biometric-based module. This module is an important subsystem of Smart grid. Such subsystem are of particular importance at module range on any abstraction level for application-specific data analysis and processing, thus allowing to leverage the underlying communication infrastructure and use and combine information generated by various devices, to produce added value across multiple environments. In a biometric-based module, there are different modalities that can be used (fingerprints, face, iris, retina, voice etc.). Among those modalities, face is considered one of the top choices because, face recognition is one of the most common human experiences: it is easy to capture at a distance and in a non-cooperative manner and, finally, face recognition technology base on selective Neural Network is fairly

* This work is supported by Russian Foundation for Basic Research (grant №14-41-0402-a)

IOP Conf. Series: Materials Science and Engineering **94** (2015) 012007 doi:10.1088/1757-899X/94/1/012007

accurate. Automatic face recognition by computer can be divided into two approaches [2], namely, content-based and face-based. In content-based approach, recognition is based on the relationship between human facial features such as eyes, mouth, nose, profile silhouettes and face boundary [3, 4, 5, 6]. The success of this approach relies highly on the accurately is difficult. Face-based approach [7, 8, 5, 9] attempts to capture and define the face as a whole. Principal Component Analysis (PCA) [7, 8, 10, 11, 12] has been proven to be an effective face-based approach. Sirovich and Kirby [10] first proposed using Karhunen-Loeve transform to represent human faces. In their method, faces are represented by a linear combination of weighted eigenvector, known as eigenfaces. Turk and Pentland [8] developed a face recognition system using PCA. However basic PCA-based methods suffer from two limitations, namely, poor discriminatory power and large computational load. Another disadvantage of PCA-based method is the high computational load in finding the eigenvectors. The computational complexity of this is *O (d^2)* where *d* is the number of pixels in

the training images which has a typical value of 128 x 128. The computational cost is beyond the power of most existing computers.

In view of the limitations in existing PCA-based approach, we proposed a new approach – applying PCA on Viola–Jones method for feature extraction. In the proposed method an image is decomposed into a number of subbands with different frequency components using the Viola–Jones method.

## 2. Analytical review of recognition methods

The Viola–Jones method is the first object detection framework to provide competitive object detection rates in real-time [2]. The main disadvantages of Viola–Jones method are robustness (very high detection rate and very low false-positive rate always); real time (for practical applications at least 2 frames per second must be processed); face detection (not recognition – the goal is to distinguish faces from non-faces).

The Viola–Jones method consists of 4 steps:

- Haar Features Selection;
- Creating Integral Image;
- Adaboost Training algorithm;
- Cascaded Classifiers.

The Viola–Jones method illustrated in Figure 1.

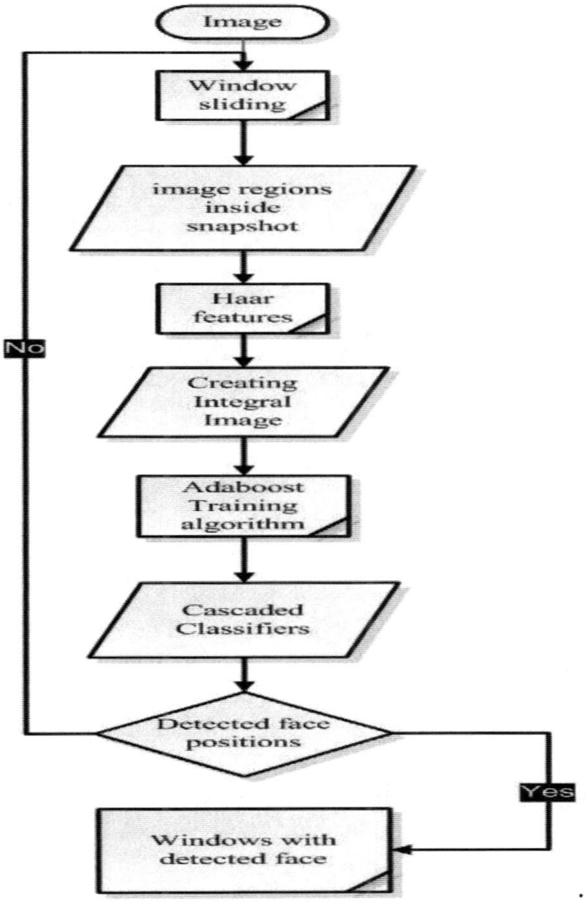


**Figure 1.** The Viola–Jones method.

The features employed by the detection framework universally involve the sums of image pixels within rectangular areas. As such, they bear some resemblance to Haar basis functions, which have been used previously in the realm of image-based object detection [3]. However, since the features used by Viola-Jones method all rely on more than one rectangular area, they are generally more complex. The value of any given feature is always simply the sum of the pixels within clear rectangles subtracted from the sum of the pixels within shaded rectangles.

The rectangular features of this sort are rather primitive when compared to alternatives such as steerable filters. Although they are sensitive to vertical and horizontal features, their feedback is considerably coarser.

PCA is used to find a low dimensional representation of data. Some important details of PCA are highlighted at [14].

Let $X = \{X_n, n = 1, \dots, N\} \in R^{dxd}$ be an ensemble of vectors. In imaging applications, they are formed by row concatenation of the image data, with dxd being the product of the width and the height of an image. The average vector in the ensemble is defined as

$$E(X) = 1/N \sum_{n-1}^{N} X_n. \quad (1)$$

After subtracting the average from each element of $X$, we get a modified ensemble of vectors,

$$\bar{X} = \{\bar{X}_n, \text{n} = 1, \dots, \text{N}\}, \quad (2)$$

where $\bar{X}_n = X_n - E(X)$.
The auto-covariance matrix *M* for the ensemble *X* is defined as

$$M = cov(\bar{X}) = E(\bar{X} \otimes \bar{X}) \quad (3)$$

where *M* is $d^2xd^2$ matrix, with elements $M(i,j) = 1/N \sum \bar{X}_n(i)\bar{X}_n(j), 1 \leq i,j \leq d^2$.
It is well known from matrix theory that the matrix *M* is positively definite (or semi-definite) and has only real nonnegative eigenvalues [13]. The eigenvectors of the matrix *M* form an orthonormal basis for $R^{dxd}$. This basis is called the *K-L* basis. Since the auto-covariance matrix for the *K-L* eigenvectors are diagonal, it follows that the coordinates of the vectors in the sample space *X* with respect to the *K-L* basis are un-correlated random variables. Let $\{Y_n, n = 1, \dots, N\}$ denote the eigenvectors and let *K* be the $d^2xd^2$ matrix which columns are the vectors $Y_1, \dots, Y_N$. The adjoint matrix of the matrix *K*, which maps the standard coordinates into *K-L* coordinates, is called the *K-L* transform. In many applications, the eigenvectors in *K* are sorted according to the eigenvalues in a descending order. In determining the *dxd* eigenvalues from *M*, we have to solve a $d^2xd^2$ matrix. Usually, *d=128* and therefore, we have to solve a 16x16 matrix to calculate the eigenvalues and eigenvectors. The computational and memory requirement of the computer systems are extremely high.

Basically, eigenface is the eigenvector obtained from PCA. This idea is first proposed by Sirovich and Kirby [10]. After that, Turk and Pentland [8] developed a face recognition system using PCA. PCA illustrated in Figure 2.

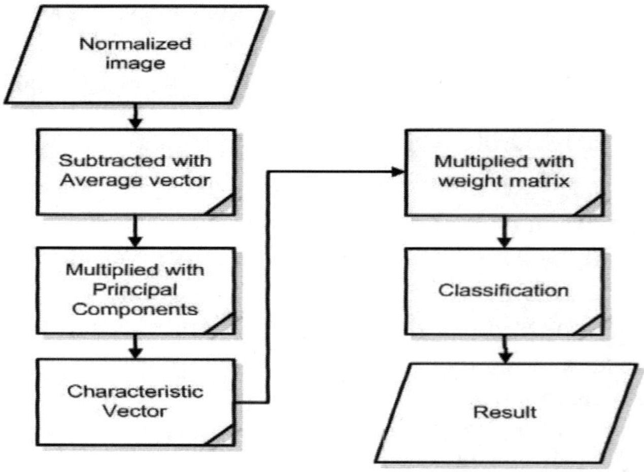


**Figure 2.** PCA block-diagram.

Multilayer Perceptron Neural Network is a good tool for classification purpose [15, 16]. In this paper, the function approximation capabilities of selective neural net [17] are exploited to recognize faces.

## 3. The proposed method

A Viola–Jones method-based PCA method is developed so as to overcome the limitation of the original PCA method; furthermore, we have utilized a selective neural net in order to carry out the classification of faces. We adopted a selective neural net [17] which is fed by the reduced input units, feature vectors generated by combination of Viola–Jones method and PCA. The combination of new Viola–Jones method-based PCA and selective neural net is illustrated in Figure 3.

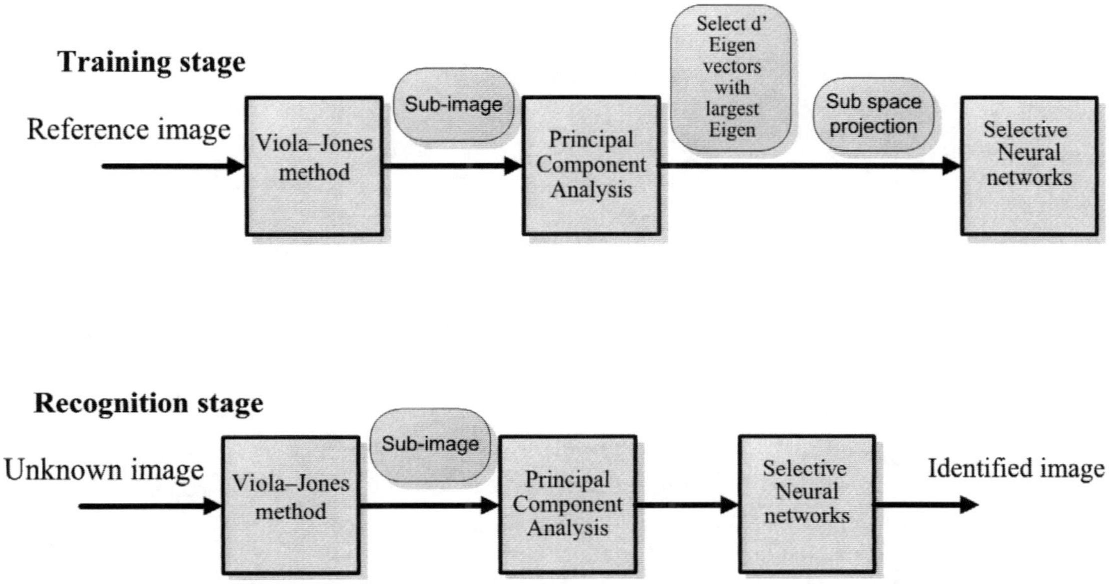


**Figure 3.** Block diagram of the proposed face recognition system.

Proposed system consists of two stages, namely training step in which the feature extraction, dimension reduction, also training of selective neural net have been performed and the recognition step identifies the unknown face image. The training stage includes the feature extraction of reference images and the adjustment of selective neural net parameters. The extracting feature identifies the representational basis for images in the domain of interest. Subsequently, the recognition stage translates the input unknown image according to the representational basis, identified in the training stage.

There are three significant steps in the training stage. In the first step, Viola–Jones method is applied to detection of faces in an image. In the next step, Principal Component Analysis (PCA) is performed on the sub-images to obtain a set of representational basis by the selection of $d'$ eigenvectors corresponding with the largest eigenvalues and sub-space projection. Finally, the feature vectors of reference images obtained by previous steps are used so as to train selective neural network using algorithm of level-by-level generating [17]. Processing in the recognition stage is similar to the training stage, except that recognition stage also incorporates steps to match the input unknown images with those reference images in the database by selective neural net. When an unknown face-image is presented to the recognition stage, Viola–Jones method and PCA are applied to transform the unknown face-image into the representational basis identified in the recognition stage, and the classification is achieved by trained selective neural net [17].

The biometric-based module carries out function of identification of the user of system. The biometric-based module's architecture consists of three components: the face detection on image, the image normalization and the face recognition. Face recognition module works on the basis of Viola-Jones method, PCA and trained selective neural net. The biometric-based module's architecture illustrated in Figure 6.

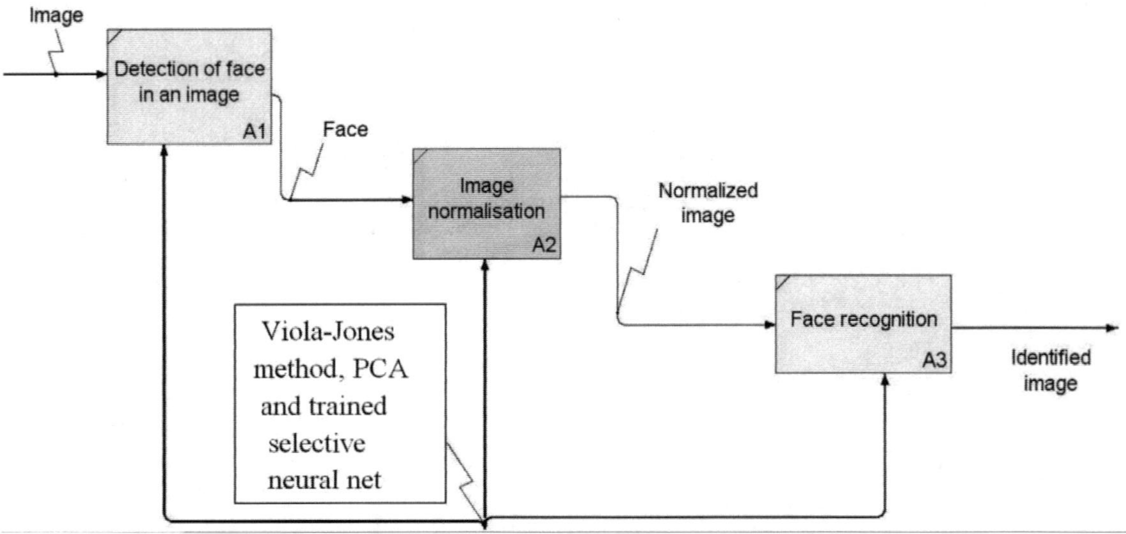


**Figure 4**. The biometric-based module's architecture.

The selective Neural Net (S NN) implemented by usage OpenCV NN.

## 4. Experimental result

To evaluate the performance of the proposed method, we used the face-images database of Yale University [18]. This database consists of 165 images (15 persons (males and females), with 11 images for each person). There are 11 images per subject, one per different facial expression or configuration. We used selective Neural Net architecture which is fed by the reduced input units, feature vectors generated by combination of Viola–Jones method and PCA. The selective neural net was trained on 22 samples (train set has 15 persons (males and females), with 2 images for each person). The experiment results are illustrated in the Table 1 and Table 2. Recognition rate on the Yale face-images database of the proposed methods is compared with recognition rate other methods [30].

**Table 1.** Performance comparison of recognition rate.

| Methods | Recognition rate (%) |
|---|---|
| IPCA (Incremental PCA) | 78,47 |
| PCA | 80,80 |
| I(2D)PCA | 81,19 |
| PCA + MLP NN | 81,27 |
| I(2D)$^2$PCA | 81,39 |
| 2DPCA | 82,05 |
| (2D)$^2$PCA | 82,13 |

| | |
|---|---|
| **PCA** | **85,71** |
| SPCA | 88,57 |
| **Viola–Jones + S NN** | **89,45** |
| WPCA | 89,52 |
| **Viola–Jones method + PCA + S NN (proposed method)** | **90,35** |
| SWPCA (Symmetrical Weighted Principal Component Analysis) | 93,33 |

Table 1 shows the good performance of the proposed method. The correct recognition accuracy with Viola–Jones method + PCA + S NN improved almost by 4,64% compared with PCA and 0,9% compared with Viola–Jones + S NN.

The proposed method provides good Recognition rate (90,35 %).

**Table 2.** The biometric-based module's result

| **Characteristics** | **Result** |
|---|---|
| The database Maximum size for a face recognition | to 500 persons |
| Recognition and identification time (at the capacity of base up to 100 records) | no more than 4 sec. |
| The maximum speed of recognition and identification | to 3 faces/sec. |

The GUI biometric-based module's result presented in Figure 5.

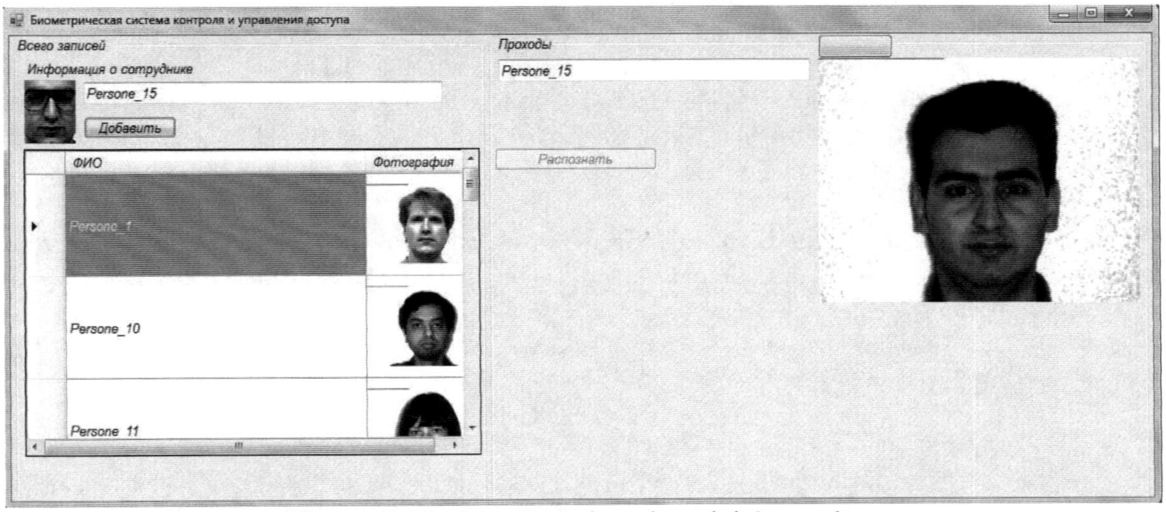


**Figure 5.** The biometric-based module's result.

The GUI biometric-based module displays image with normalized histogram (at the left), identified data (name person), input image and face-images database users of the module.

## 5. Conclusions

Within Smart Grid concept the flexible biometric-based module base on Viola–Jones method, Principal Component Analysis and selective Neural Network is developed. The formed as the selective Neural Network the biometric-based module uses the proposed method which includes three main stages: preliminary processing of the image, face localization and face recognition. Experiments on the Yale face database show that (i) selective Neural Network exhibit promising classification capability for face detection, recognition problems; and (ii) the proposed biometric-based module achieves near real-time face detection, recognition speed and the competitive performance, as compared to some existing subspaces-based methods.

The biometric-based module has flexible architecture and easily implements in Smart Grid technology. Experiments show that the biometric-based module provides good effectiveness and secure for Smart Grid technology.

## References

[1] McCaugherty Dan 2014 The IEEE Computer Society Smart Grid Vision Project, Technological Concepts, *Proc. IEEE Conf. Computer vision and Pattern Recognition, November* 84-91

[2] Chow G and Li X 2010 Towards a system for automatic facial feature detection *Pattern Recognition* vol.26 **12** 1739-1755.

[3] Goudail F, Lange E, Iwamoto T, Kyuma K and Otsu N 1995 Face recognition system using local autocorrelations and multiscale integration *IEEE Trans. PAMI* vol. 18 **10** 1024-1028

[4] Lam K M and Yan H 1996 Locating and extracting the eye in human face images *Pattern Recognition* vol. 29 **5** 771-779

[5] Valentin D, Abdi H, O'Toole A J and Cottrell G W 1994 Connectionist models of face processing: A Survey, Pattern Recognition vol.27 1209-1230

[6] Yuille A L, Hallinan P W and Cohen D S 1992 Feature extraction from faces using deformable templates, Int. J. of Computer Vision vol.8 **2** 99-111

[7] Kirby M and Sirovich L 1990 Application of the Karhunen-Loeve procedure for the characterization of human faces *IEEE Trans. PAMI.* vol.12 103-108

[8] Turk M and Pentland A 2001 Eigenfaces for recognition *J. Cognitive Neuroscience* vol.3 71-86

[9] Wickerhauser M V 2014 Large-rank approximate component analysis with wavelets for signal feature discrimination and the inversion of complicated maps *J. Chemical Information and Computer Sciences* vol.34 **5** 1036-1046

[10] Sirovich L and Kirby M 2007 Low-dimensional procedure for the characterization of human faces *J. Opt. Soc. Am. A* vol.4 **3** 519-524

[11] Swets D L and Weng J J 2006 Using discriminant eigenfeatures for image retrieval *IEEE Trans. PAMI.* vol.18 **8** 831-836

[12] O'Toole A J, Abdi H, Deffenbacher K A and Valentin D 2013 A low-dimensional representation of faces in the higher dimensions of the space *J. Opt. Soc. Am. A* vol.10 405-411

[13] Jain A K 2009 Fundamentals of digital image processing *Prentice Hall* pp.163-175

[14] Moghaddam B, Wahid W and Pentland A 2008 Beyond eigenfaces: Probabilistic matching for face recognition *Proceeding of face and gesture recognition* pp. 30

–35

[15] Wasserman A I 2014 *Neural Computing: Theory and Practice – New York: Van Nostrand Reinhold*

[16] Golovko V, Gladyschuk V 1999 Recirculation Neural Network Training for Image Processing *Advanced Computer Systems* pp. 73-78.

[17] Engel E A 2012 Solving control problems, decision making and information processing by fuzzy selective neural network// Information Technology, 2

[18] Yale face database: http://cvc.yale.edu/projects/yalefaces/yalefaces.html

[19] Fernandes S and Bala J 2013 Performance Analysis of PCA-based and LDA-based Algorithms for Face Recognition, *Coimbatore, India, June*

# ENERGY-SAVING TECHNOLOGY OF VECTOR CONTROLLED INDUCTION MOTOR BASED ON THE ADAPTIVE NEURO-CONTROLLER*

**E Engel[1], I V Kovalev[2], D Karandeev [3]**
[1] Associate Professor, First Russian doctoral degree in computer sciences, Katanov State University of Khakassia, Abakan, RF,
[2] Professor, Second Russian doctoral degree in computer sciences, Siberian State Aerospace University, Krasnoyarsk, RF,
[3] Undergraduate student, Katanov State University of Khakassia, Abakan, RF

E-mail: Angel@khsu.ru

**Abstract.** The ongoing evolution of the power system towards a Smart Grid implies an important role of intelligent technologies, but poses strict requirements on their control schemes to preserve stability and controllability. This paper presents the adaptive neuro-controller for the vector control of induction motor within Smart Gird. The validity and effectiveness of the proposed energy-saving technology of vector controlled induction motor based on adaptive neuro-controller are verified by simulation results at different operating conditions over a wide speed range of induction motor.

## 1. Introduction

In the modern industrial sector induction motors occupy a vital role. As compared to DC drives induction motors play an important role in the safe and efficient operations of industrial plants because of its high efficiency, simplicity of design as well as reliability and relatively low cost and maintenance. There are many different ways to drive an induction motor such as scalar control, vector control, direct torque control and etc. The present paper reviews the main concept of Field Orientated Control (FOC) or vector control of induction motors [1]. In this work we focus on the vector control of induction motor within Smart Gird. The induction motor drive is a dynamic nonlinear system with time varying uncertainties. The ongoing evolution of the nonlinear system of induction motor towards a Smart Grid [2] implies an important role of intelligent technologies, but poses strict requirements on their control schemes to preserve stability and controllability. Nowadays around 70% of electric power is consumed by electric drives. In the industrial sector is increasing the requirements for energy-saving technology. The proportional integral (PI) controllers are widely used in the field-orientated control of induction motor. But it takes even experienced engineers time and effort to regulate the PI coefficients to guarantee good performance. Usually PI controller is tuned only at particular operation point. If the operating point changes then it not operate properly. Therefore, automatic design methods utilizing machine learning techniques such as neuro-controller are a promising alternative as compared to the PI controller. This paper presents the energy-saving technology of vector controlled induction motor based on adaptive neuro-controller within Smart Gird. In this work, the adaptive neuro-controller is designed in two

* This work is supported by Russian Foundation for Basic Research (grant №14-41-0402-a)

steps. The first step objective is dimensionality reduction, which is developed through Principal Component Analysis. The second step objective is to generate control signal, which is developed using selective neural network [3].

## 2. Mathematical model of the induction motor

The mathematical model of the induction motor used in our work consists of space vector PWM voltage source inverter, induction motor, direct flux and the torque control. Within field-oriented control an induction motor model was established in the *d-q* motor reference frame (without saturation) concept. Figure 1 shows the equivalent circuit which used for obtaining the mathematical model of the induction motor.

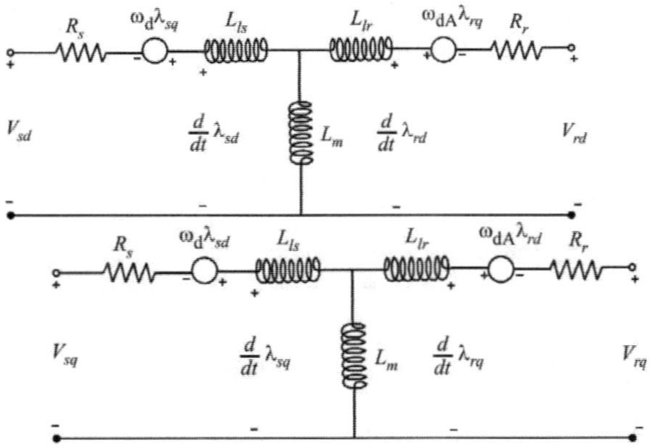

**Figure 1**. Equivalent circuit of induction motor in *d-q* frame.

The dynamics of the induction motor in *d-q* coordinates established in a rotor flux oriented reference frame is given by the following equations [1]:

$$V_{ds}^e = R_s i_{ds}^e + p\lambda_{ds}^e - \omega_e \lambda_{qs}^e$$
$$V_{qs}^e = R_s i_{qs}^e + p\lambda_{qs}^e + \omega_e \lambda_{ds}^e$$

$$0 = R_r i_{dr}^e + p\lambda_{dr}^e - (\omega_e - \omega_r)\lambda_{qr}^e$$
$$0 = R_r i_{qr}^e + p\lambda_{qr}^e + (\omega_e - \omega_r)\lambda_{dr}^e$$

where $\lambda_{ds}^e = L_s i_{ds}^e + L_m i_{dr}^e$, $\lambda_{qs}^e = L_s i_{qs}^e + L_m i_{qr}^e$, $\lambda_{dr}^e = L_r i_{dr}^e + L_m i_{ds}^e$, $\lambda_{qr}^e = L_r i_{qr}^e + L_m i_{qs}^e$.

The electromagnetic torque expressed in terms of inductances is described by equations [2]:

$$T_e = \frac{3}{2}\frac{P}{2} L_m (i_{qs}^e i_{dr}^e - i_{ds}^e i_{qr}^e)$$
$$\frac{d\theta_r}{dt} = \omega_r ,$$
$$T_e = j_m \frac{d\omega_r}{dt} + B_m \omega_r + T_l$$

where $V_{ds}^e, V_{qs}^e$ are *d-q* axis stator voltages respectively; $i_{ds}^e, i_{dr}^e, i_{qs}^e, i_{qr}^e$ are *d-q* axis stator currents and *d-q* axis rotor currents respectively; $R_s$, $R_r$ are the stator and rotor resistance per phase respectively; $L_s$, $L_r$. $L_m$ are the self inductances of the stator and rotor and the mutual inductance respectively; *P* is the number of poles; *p* is the differentiation operator (*d/dt*); $\omega_e$, $\omega_r$ the speed of the rotating magnetic field and the rotor speed respectively; $T_e$, $T_l$ are the electromagnetic developed torque and the load torque respectively; $J_m$ is the rotor inertia; $B_m$ is the rotor damping coefficient and $\theta_r$ is the rotor position.

## 3. Field-Orientated Control of induction motor

The field-oriented control (FOC) was initially developed by Blaschke (1971-1973) [4]. Vector control presents following benefits: speed control over a wide range, precise speed regulation, fast dynamic response, and operation above base speed. In spite of complexity, vector control plays a key role for induction drives, replacing scalar control. The FOC's key idea is usage projections which transform a three phase time and speed dependent system into a two coordinate time invariant system. This handles instantaneous electrical quantities and makes the control accurate in every working operation (steady state and transient) and independent of the limited bandwidth mathematical model. Field orientated controlled machines had two constants as input: the torque (aligned with the *q*) and the flux (aligned with *d*). Knowledge of the rotor flux position is the core of the FOC. Field oriented controlled induction machines obtain every DC machine advantage: instantaneous control of the separate quantities allowing accurate transient and steady state management [5]. Figure 2 shows the core principle of indirect vector control.

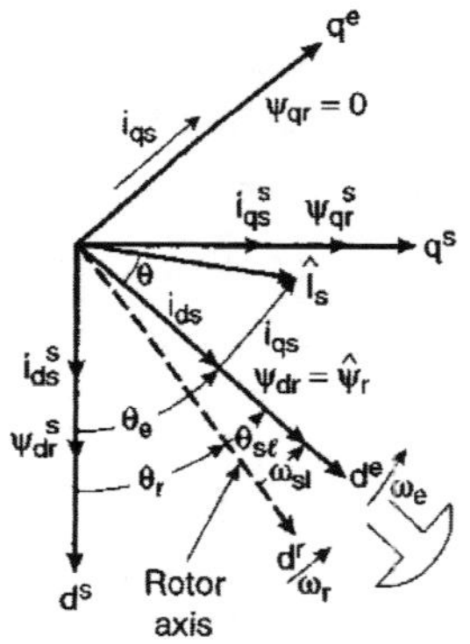


**Figure 2.** Phasor diagram of indirect vector controlled induction machine.

The axes are fixed on the stator, but the $d^r$-$q^r$ axes, which are fixed on the rotor, are $d^s$-$q^s$ moving at speed $\omega_r$. Synchronously rotating axes $d^e$-$q^e$ are rotating ahead of the $d^r$-$q^r$ axes by the positive slip angle $\theta_{sl}$ corresponding to slip frequency $\omega_{sl}$. The induction motor always relies on a small difference in speed between the stator rotating magnetic field and the rotor shaft speed called slip to induce rotor current in the rotor AC winding [2]. If the speed controller is intelligently designed, it will have the ability to minimize de-tuning effects and the drive performance will be very robust [3].

## 3. The adaptive neuro-controller for vector controlled induction motor drive

In order to improve effectiveness of indirect vector controlled induction machine we designed the adaptive neuro-controller. We created the adaptive neuro-controller using input–output data

$$(X, i_q) \tag{1}$$

generated from the Simulink model of vector controlled induction machine (Fig. 3) [7].

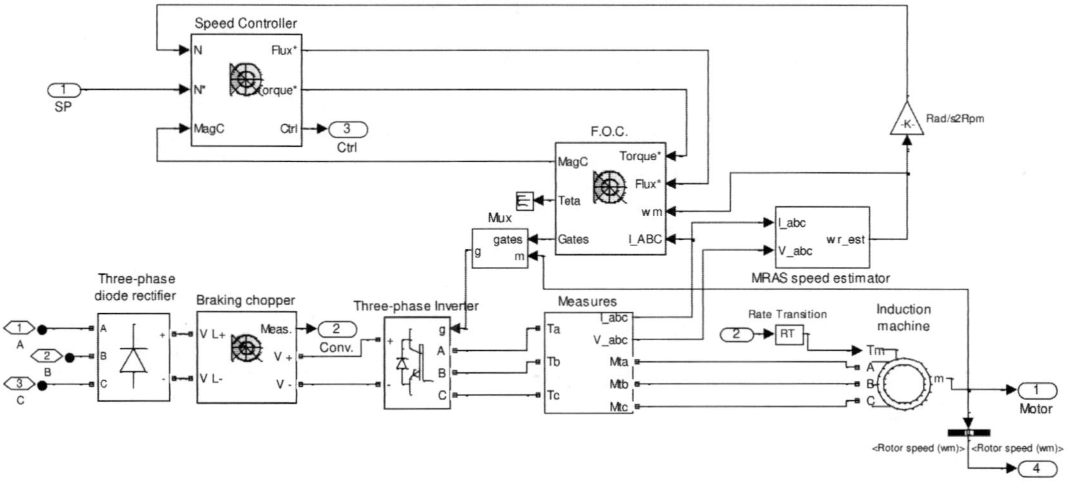


**Figure 3.** Simulink model of indirect vector controlled induction machine. [7]

In this work, the adaptive neuro-controller is designed in two steps. The first step objective is dimensionality reduction of data (1) ($X \in R^{21}$) from 21 to 5, which is developed through Principal Component Analysis. This step forms training data set $(x, i_q)$ for the selective neural network ($x \subset X$, $x \in R^5$). The statistical features $x = (i_{sa}, i_{sb}, i_{sc}$, electromagnetic moment and speed of rotor) extracted from the data (1) are used as inputs to selective neural network. The second step objective is to generate control signal $i_q$, which is developed using the selective neural network [3]. The selective neural networks have capable of machine learning as well as pattern recognition thanks to their adaptive nature. The fulfilling of the structure and learning algorithm of the selective neural network are explained as follows [3].
The proposed scheme with adaptive neuro-controller was implemented using MATLAB/ Simulink software.

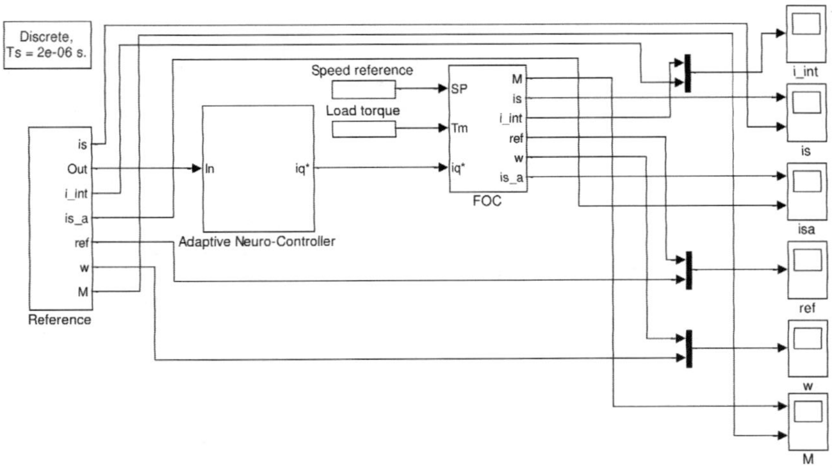


**Figure 4.** Simulink model of vector controlled induction motor based on the adaptive neuro-

controller.

In order to verify the performance and effectiveness of proposed control scheme with adaptive neuro-controller comparative simulations was conducted in MATLAB/Simulink. (Fig.4)

## 4. Simulation results

Simulation was conducted among Matlab/Simulink package. In this paper, the pattern of the reference model $i_q$ that was used to train of the proposed controller for the vector controlled induction motor and the adaptive neuro-controller's control signals are shown in Fig 5. The parameters for the induction motor are 460V, 60Hz, Poles=4, Rs=14.85e-3Ω, Rr=9.295e-3Ω, Ls=0.3027e-3H, Lr=0.3027e-3H, Lm=10.46e-3H, J=3.1kg-m2.

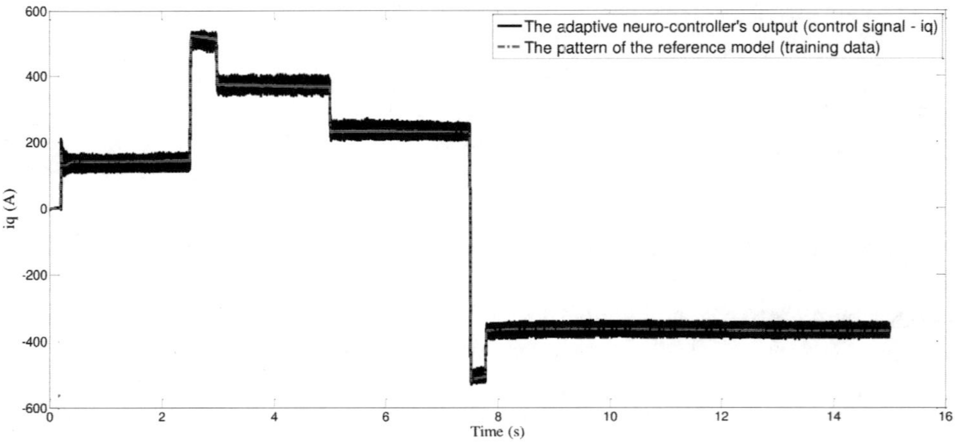


**Figure 5.** The pattern of the reference model $i_q$ (training data) and the adaptive neuro-controller's output.

In our work we compared PI and the adaptive neuro-controller for speed control of vector control induction motor drive. Figure 6 shows that the proposed scheme with adaptive neuro-controller provides energy saving due reduction of losses in the windings of the stator current as compared with PI controller.

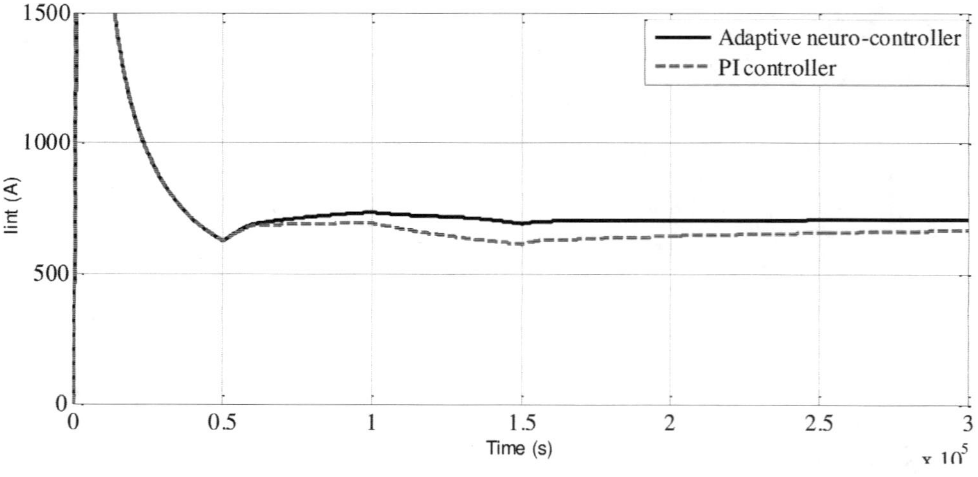

**Figure 6.** Integral estimation of currents stator $i_s - I_{int}$ (A).

The aforementioned dynamic simulation results performance and robustness of the proposed energy-saving technology of induction motor based on adaptive neuro-controller are confirmed as compared to PI controller. Also the energy-saving technology of vector controlled induction motor based on adaptive neuro-controller demonstrates the good quality and robustness in the system dynamic response and reduction in the steady-state and transient motor ripple torque.

## 5.Conclusions

The validity and effectiveness of the proposed energy-saving technology of vector controlled induction motor based on adaptive neuro-controller are verified by simulation results at different operating conditions over a wide speed range of induction motor.
Based on simulation comparative results, the following conclusions are made:

- the proposed scheme with adaptive neuro-controller has better performance over PI controller;
- the proposed adaptive neuro-controller provides energy saving due reduction of losses in the windings of the stator current as compared with PI controller;
- the proposed energy-saving technology of vector controlled induction motor based on adaptive neuro-controller is more robust during the load changes and eliminates the transients during sudden changes in speed;
- the performance of the adaptive neuro-controller is improved by reducing the ripple torque as compared with PI controller;
- the simulation results under random perturbation identified the obvious validity of the proposed energy-saving technology of vector controlled induction motor based on adaptive neuro-controller.

Thus it is found that the energy-saving technology of vector controlled induction motor based on adaptive neuro-controller can be effectively used to control induction motor drive in practice.

## References.

[1] Vas P 1988 Sensorless vector and direct torque control *Oxford University Press.*
[2] Vandoorn T L, Zwaenpoel B, Kooning J D M De , Meersman B, Vandevelde L 2011 Smart microgrids and virtual power plants in a hierarchical control structure *Proc. Of the 2nd IEEE PES Int. Conference and Exhibition on Innovative Smart Grid Technologies, ISGT Europe 2011, Manchester, UK, 5–7 December 2011* pp.7.
[3] Engel E A 2012 Solving control problems, decision making and information processing by fuzzy selective neural network *Information Technology* **2**.
[4] Leonhard W 1985 Control of electrical drives *Springer-Verlag, Berling*.
[5] Blaschke F 1971 A new method for the structural decoupling of A.C. induction machines, *Conf. Rec. IFAC, Duesseldorf, Germany*, Oct. 1971, pp. 1–15.
[6] Popescu M 2000 Induction motor modeling for vector control purposes *Helsinki University of Technology, Laboratory of Electromechanics, Report, Espoo* 144 p.
[7] http://www.mathworks.com/ (date of access 01.03.2014).
[8] Blaschke F 1972 The principle of fields-orientation as applied to the Transvector closed-loop control system for rotating-field machines *Siemens Review* **34** pp.217-220.

# INTELLIGENT CONTROL OF NON-LINEAR DYNAMICAL SYSTEM BASED ON THE ADAPTIVE NEURO-CONTROLLER*

**E Engel[1], I V Kovalev[2], V Kobezhicov[3]**
[1] Associate Professor, First Russian doctoral degree in computer sciences, Katanov State University of Khakassia, Abakan, RF,
[2] Professor, Second Russian doctoral degree in computer sciences, Siberian State Aerospace University, Krasnoyarsk, RF,
[3] Student, Katanov State University of Khakassia, Abakan, RF

E-mail: Angel@khsu.ru

**Abstract.** This paper presents an adaptive neuro-controller for intelligent control of non-linear dynamical system. The formed as the fuzzy selective neural net the adaptive neuro-controller on the base of system's state, creates the effective control signal under random perturbations. The validity and advantages of the proposed adaptive neuro-controller are demonstrated by numerical simulations. The simulation results show that the proposed controller scheme achieves real-time control speed and the competitive performance, as compared to PID, fuzzy logic controllers.

## 1. Introduction

Non-linear dynamical systems are commonly suffering from restrictions imposed by the uncertainty of the environment and therefore unstable dynamics when designing control systems to provide stability, disturbance attenuation, and reference signal tracking. Within the research literature, a whole array of differing control strategies is proposed to deal with the control of the linearized system problem. One of the most common control strategies is Proportional-Integral-Derivative (PID) controller due to its simplicity and applicability [1]. But the PID controllers for non-linear dynamical system are often designed by hand, requiring extensive analysis of the system and dynamics. This process is generally difficult because it is hard to anticipate all operating conditions. The controller must coordinate the non-linear dynamical system properly, generating robust behavior to negotiate different terrains effectively while maintaining stability. Moreover, the non-linear dynamical system should be robust to different environmental conditions, wear and tear, and even failure to reliably complete its mission. Therefore, automatic design methods utilizing intelligent techniques such as neural network (NN) and fuzzy logic are a promising alternative [2].
Monitoring and fault detection play an important role for non-linear dynamical system due to the increasing demands on fault tolerant real-time applications. This fault tolerant behavior of non-linear dynamical system refers to the possibility to autonomously detect faults early before they result in catastrophic failures as well as to the capability to continue operating after a fault has occurred by switching to a safe state. However, fault detection for non-linear

* This work is supported by Russian Foundation for Basic Research (grant №14-41-0402-a)

dynamical system is a complex problem because of the large space of possible faults (e.g. non-linear system's sensors, actuators, the uncertainty of the environment). Non-linear dynamical object are best modeled as hybrid systems since their behaviors result from the interaction between continuous and discrete dynamics. Several methods have been developed to deal with the monitoring of such systems. The most important approaches are these that combine the basic model, which are differential equations, with the intelligent models, which are neural network and fuzzy logic. This paper presents an adaptive neuro-controller for intelligent control of non-linear dynamical system.

## 2. Adaptive neuro-controller on the base of the fuzzy selective neural net

A large body of literature on neuro-adaptive control exists on nonlinear systems of the form $\dot{x}=f(x)+g(x)u$, where the neural nets are used to approximate $f(x)$ and $g(x)$ [3]. Adaptive control methods such as linearization have been shown to be very effective for the control of a broad class of systems [3]. In contrast, in this paper, the function approximation capabilities of fuzzy selective neural net [4] are exploited to approximate a nonlinear control law. The adaptive neuro-controller is capable of handling uncertainties in both the system parameters and the environment. The formed as the fuzzy selective neural net the adaptive neuro-controller creates the effective control signal according identified system's state under random perturbations (Fig. 1). The fuzzy selective neural net trained base on data

$$Z^i=(s_1^i,\ s_2^i,\ \ldots,\ s_r^i,\ u_1^i,\ u_2^i,\ldots\ u_b^i,\ \Omega_1^i,\ \Omega_2^i,\ \ldots,\ \Omega_q^i,\ Y), \qquad (1)$$

where $i\in\{1,.., m\}$ – time, $Y$ – response of the nonlinear system; $s$ – input signal of the nonlinear system; $u$ – control signal; $\Omega$ – influences of environment.

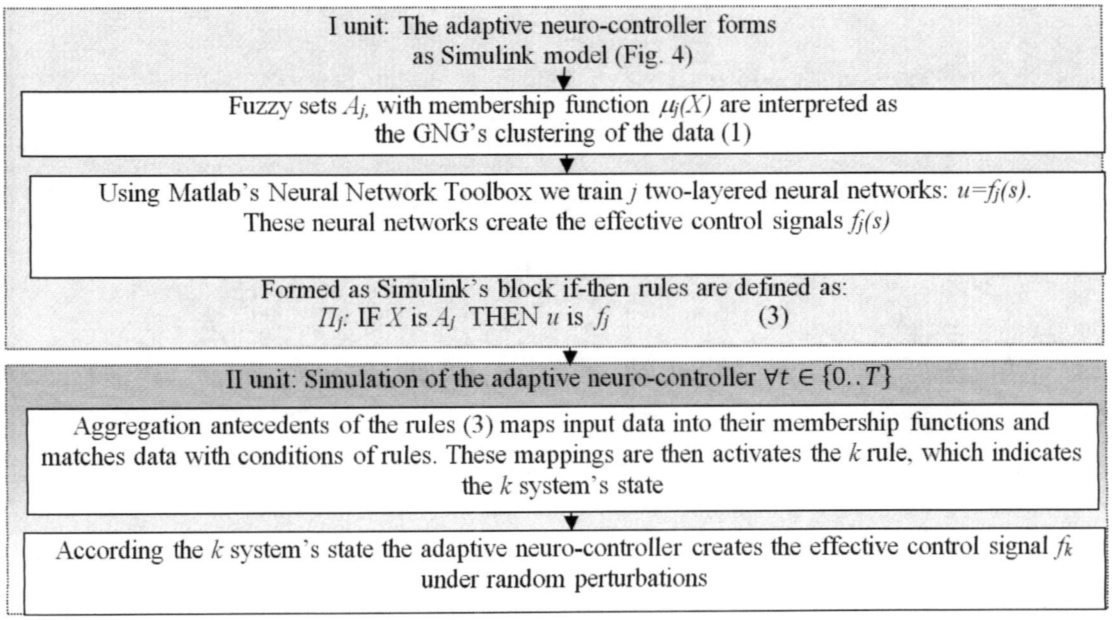


**Figure 1.** The adaptive neuro-controller.

First, in order to obtain the dynamics of the system the fuzzy sets $A_j$ with membership function $\mu_j(X)$ formed base on GNG's clustering (Growing Neural Gas [5]) of the data (1). Second, for each $j$ an identifier is constructed by a two-layer feed forward neural network.

Third, an adaptive neuro-controller is carefully designed to correctly tackle the control task under uncertainty of the nonlinear system and of the environment.

## 3. Dynamic modeling of the mobile robot

In this paper a mobile robot prototype with four DC motor actuators was considered as an example of non-linear dynamical system. The drive and steering subsystem is included two independent wheels and motor [6]. The robot navigates by changing the relative rate of rotation of its wheels motor. The physical parameters of the motors are shown in Table 1.

**Table 1.** The physical parameters of the motors.

| Parameters | values |
|---|---|
| Moment of inertia of the rotor (J) | 42.6e-6Kgm2 |
| Viscous friction coefficient (b) | 47.3e-6Nms |
| Torque constant ($k_t$) | 14.7e-3$Nm/Amp$ |
| Back emf ($k_e$) | 14.7e-3 $V.s/$rad |
| Terminal resistance (R) | 4.67$ohm$ |
| Electric inductance (L) | 170$e$ -3 $H$ |

The behavior of the motor speed for a given voltage is derived from physics law described in Laplace domain by the open-loop transfer function

$$\frac{\dot{\ }(s)}{V(s)} = \frac{\frac{K_t}{JL}}{s^2+\frac{(JR+bL)}{JL}s+\frac{bR+K_eK_t}{JL}} \qquad (2)$$

where voltage $V(s)$ is input and shaft speed $\omega = \dot{\theta}(s)$ is output [7]. The relation between the reference speed $\dot{\theta}_{ref}$ and the output speed $\dot{\theta}_{out}$ is given by the closed loop transfer function

$$\frac{\dot{\theta}_{out}}{\dot{\theta}_{ref}} = \frac{K.\frac{K_t}{JL}}{S^2+\frac{(JR+bL)}{JL}S+\frac{K.K_t+bR+k_eK_t}{JL}}$$

where $K$ is a constant gain.
The state space representation of the robot's dynamic as follows

$$\begin{bmatrix}\frac{di}{dt}\\ \ddot{\theta}\end{bmatrix} = \begin{bmatrix}-\frac{R}{L} & -\frac{K_e}{L}\\ \frac{K_t}{J} & -\frac{b}{J}\end{bmatrix}\begin{bmatrix}i\\ \dot{\theta}\end{bmatrix} + \begin{bmatrix}\frac{1}{L}\\ 0\end{bmatrix} V(t) \text{and } \dot{\ } = [0\ 1]\begin{bmatrix}i\\ \dot{\theta}\end{bmatrix} \qquad (3)$$

was used to confirm that the system is stable, controllable and observable.

## 4. PID controller design

The general transfer function for a PID controller in Laplace domain can be written as follows

$$PID(s) = K_p + \frac{K_I}{s} + K_D\, s = \frac{K_D\, s^2 + K_p\, s + K_I}{s} \qquad (4)$$

where $K_p$ is the proportional gain, $K_D$ is the derivative gain and $K_I$ is the integrator gain. The PID gains were choose by hand and were then tuned by simulation with following values: $K_P$=0.68, $K_I$=0.6, $K_D$=0.01.
The open loop and closed loop transfer function of the system with PID controller is given by (5) and (6) respectively:

$$G_{open\_loop\_PID}(s) = \frac{24.15\, s^2 + 1389\, s + 1213}{s^3 + 28.58\, s^2 + 60.34\, s} \quad (5)$$

$$G_{closed\_loop\_PID}(s) = \frac{24.15\, s^2 + 1389\, s + 1213}{s^3 + 52.74\, s^2 + 1450 s + 1213} \quad (6)$$

This PID controller was designed by hand. Therefore, automatic design methods utilizing intelligent techniques such as neural network and fuzzy logic are a promising alternative.

## 5. Intelligent control

An intelligent approach is proposed to address the weaknesses of the PID approach. The adaptive neuro-controller is carefully designed to correctly tackle the control of non-linear dynamical system task under uncertainty of the environment.

### *5.1. Fuzzy logic controller design.*

A typical structure of a fuzzy logic controller is shown in Fig. 2.

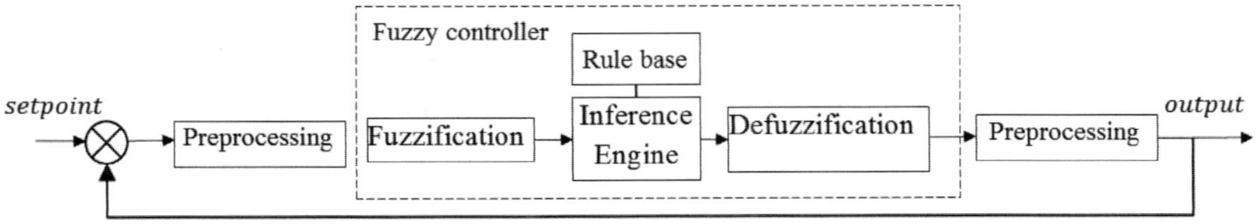


**Figure 2.** Fuzzy logic controller.

In our work, knowledge was borrowed from proportional integral control error and change of error to define fuzzy membership functions. The inputs that were in the form of crisp values generated from feedback error (e) and change of error (de) were conditioned in terms of multiplying by constant gains before entering into the main control block. The fuzzification block converts input signals into appropriate way to membership functions, invokes appropriate rules, generates a result for each rule, and combines the results of those rules. The Mamdani-type inference engine combined the output state and assigned a membership value at the truth level of the premise. The truth values are then defuzzified using a discrete centroid computation. The post processing block then transforms these crisp values into control signals. The fuzzy logic controller's rules had triangular membership function and were implemented in the simulation.
These rules provide control signals based on several if-then statements about (e) and (de), i.e., if the error is equal Negative Big (NB) and change of error is equal to negative medium (NM), then the change in control (c) is positive big (PB). The numbers of fuzzy logic controller's rules were determined based on experiment and tuning of the system.

IOP Conf. Series: Materials Science and Engineering **94** (2015) 012009 doi:10.1088/1757-899X/94/1/012009

*5.2. The adaptive neuro-controller design*

With onboard resources often limited, this paper considers the development of an effective control method that remains easy to implement. The formed as the fuzzy selective neural net the adaptive neuro-controller on the base of system's state, creates the effective control signal under random perturbations (Fig. 1). First, in order to obtain the dynamics of the system, the fuzzy sets $A_j$ with membership function $\mu_j(X)$ formed base on GNG's clustering of the data (1), $j \in \{1, 2, 3\}$. Second, for each $j$ an identifier is constructed by a two-layer feed forward neural network. Third, an adaptive neuro-controller is carefully designed to correctly tackle the control task under uncertainty of the robot and of the environment. To make the neural network become adaptive, it needs to have some idea on how the actual robot's behavior is differing from its expected behavior, so that the controller can recalibrate its behavior intelligently during run time, and try to eliminate the constant tracking error. Hence the input signal of the neuro-controller will be non-zero, and it will give useful feedback for telling the controller how to adapt to the dynamically changing robot's conditions. This control approach does provide a more intelligent method of implementing the control signal.

## 6. Simulation and results

All the simulations for this study are implemented in MATLAB, Simulink. The Simulink block diagram for the implementation is shown in Figure 4.

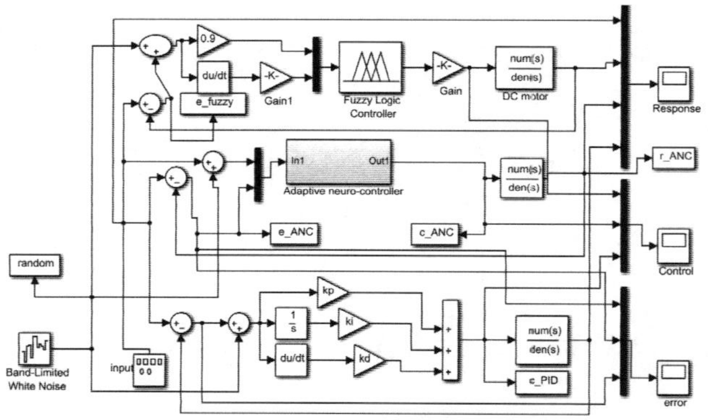


**Figure 4.** Simulation diagram of controllers for DC motor speed.

To illustrate the benefits of the newly proposed intelligent control, the numerical examples from the previous section 4 is revisited. The true benefits of the proposed controller are best demonstrated through a simulation study.

In this comparison study, the performance of the adaptive neuro-controller is compared against fuzzy logic and standard PID controllers, under the same conditions. Fig. 5 to 11 shows the simulation's results.

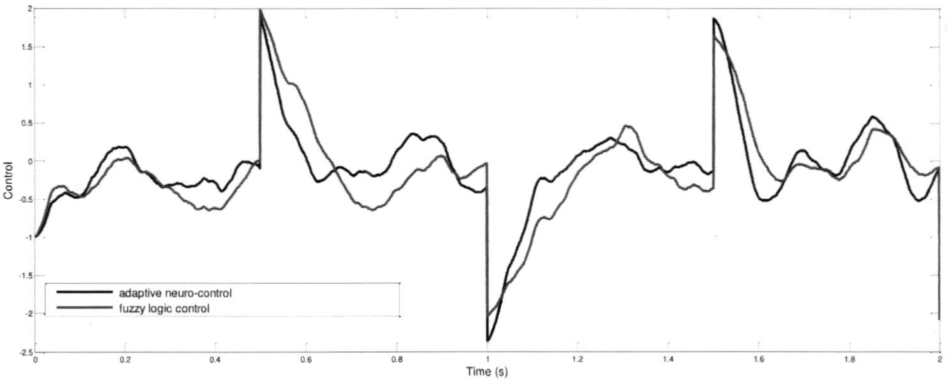


**Figure 5.** Plot of control signal for adaptive neuro-controller and fuzzy logic controller for step input.

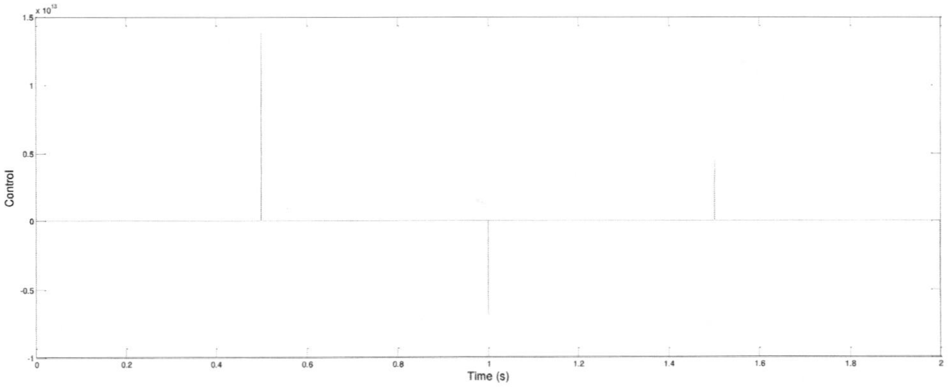


**Figure 6.** Plot of control signal for PID controller for step input.

It can be seen that the PID controller takes a huge numerical value of control signal in the step changes (Fig. 6), while the other two controllers remain in the range of reasonable values of [-2, 2] (Fig. 5). One of the result curves (Fig. 5) is for adaptive neuro-controller, another is fuzzy logic controller

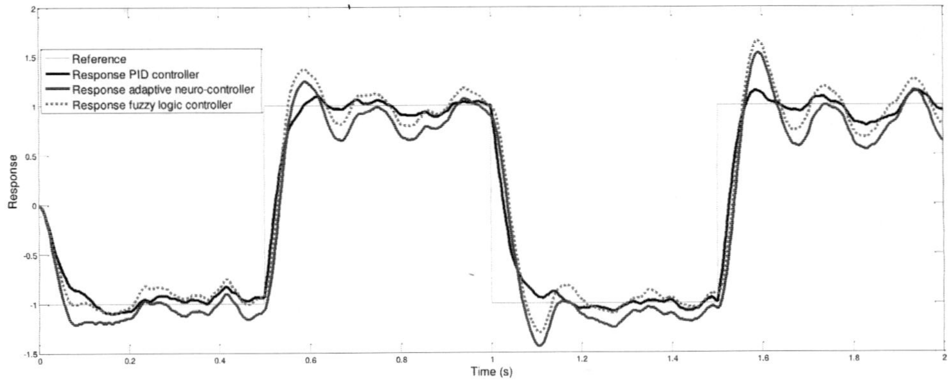


**Figure 7.** Plot of step response and performance comparison for the designed controllers.

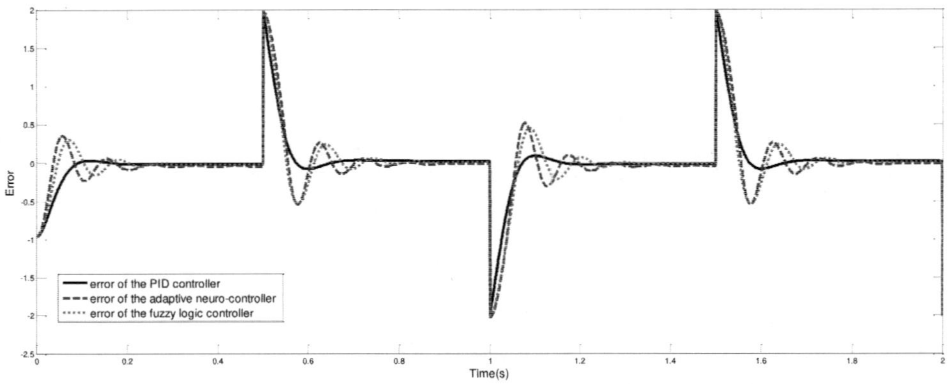


**Figure 8.** Plot of errors to the controllers responses.

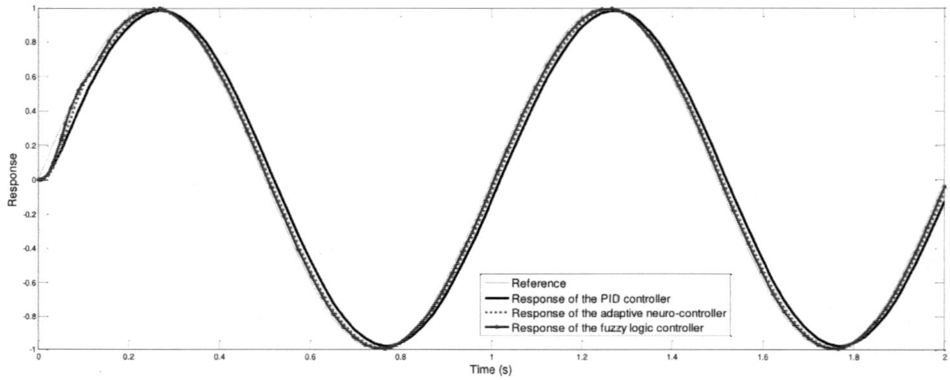


**Figure 9.** Plot of sinusoidal response and performance comparison for the designed controllers.

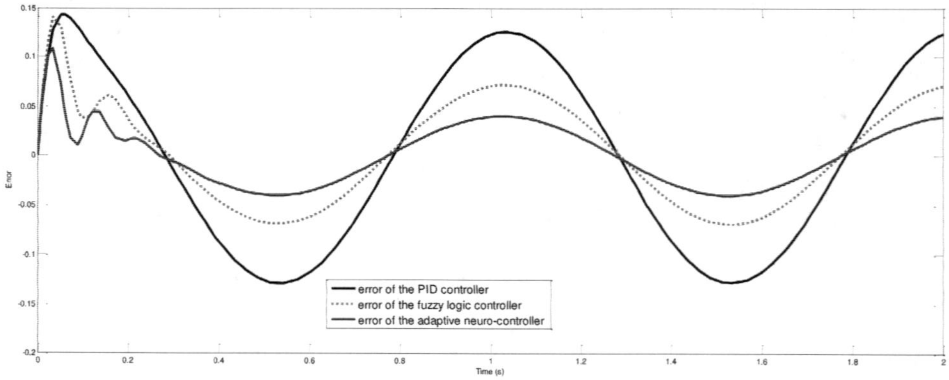


**Figure 10.** Plot of errors to the controllers responses for sinusoidal input.

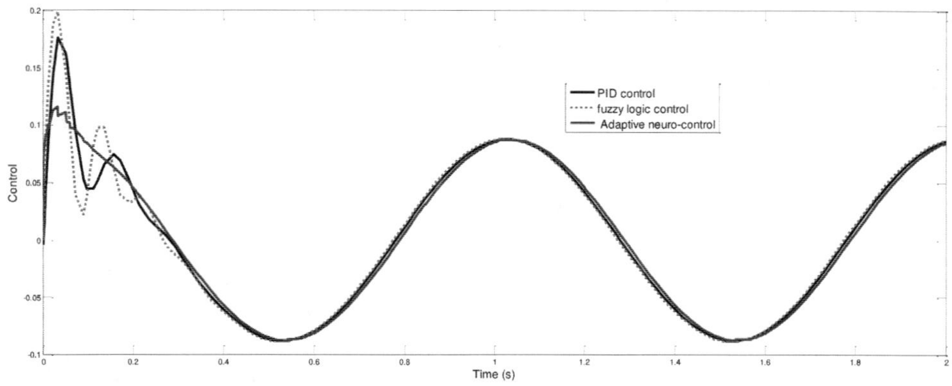


**Figure 11.** Plot of control signal for neural and fuzzy logic controllers for step input.

The control signal of the adaptive neuro-controller is more robust and less intensive in compare with the fuzzy logic controller (Fig. 5, 11). The tolerance of an adaptive neuro-controller to noise explained by two factors: the ability of the model to have similar responses for patterns contaminated with different intensities of noise and the resilience to noise of the low similarity of the responses for patterns of different system's state.

Use of the adaptive neuro-controller provides a more suitable approach to the control problem, with the pointing accuracy. Extensive simulation studies on Simulink model have been carried out on different initial conditions, different disturbance profiles and variation in robot parameters. It shows consistent performance has been achieved for the proposed adaptive neuro-controller with good stability and robustness.

## 7. Conclusions

It is shown that the adaptive neuro-controller is robust to system uncertainties. Unlike popular approaches to nonlinear control, an adaptive neuro-controller is used to approximate the control law, and not the system nonlinearities, which makes it suitable to handle a wide range of nonlinearities. Simulations on the intelligent control system for wheel speed and steering of mobile robot demonstrate the effectiveness of the adaptive neuro-controller.

## References

[1] Ang K H, Chong G and Li Y 2005 PID control system analysis: Design and technology. *IEEE T. Contr. Syst. T.* **13(4)** pp. 559-576.

[2] Makarov I M, Lokhin V M, Manko S V, Romanov M P, Sitnikov M S 2013 Stability of intelligent systems of automatic control *Information Technology* **2**

[3] Kovalev I, Zelenkov P, Ognerubov S, Bahmareva K, Denisova E 2015 The efficiency analysis of the automated plants *IOP Conference Series: Materials Science and Engineering* **70 (1)**

[4] Omidvar O M, Elliott D L 1997 Neural Systems for Control *Elsevier*

[5] Engel E A 2012 Solving control problems, decision making and information processing by fuzzy selective neural network *Information Technology* **2**

[6] Isaac J Sledge 2008 Growing Neural Gas for Temporal Clustering *IEEE*

[7] Lucas G W 2001 A Tutorial and Elementary Trajectory Model for the Differential Steering System of Robot Wheel Actuators Retrieved from: http://rossum.sourceforge.net/papers/DiffSteer/.

[8] Dorf R C and Bishop R H 2001 Modern Control Systems *9th Ed., Prentice-Hall, Inc., Upper Saddle River, NJ*

# ADAPTATION ALGORITHMS FOR SATELLITE COMMUNICATION SYSTEMS EQUIPPED WITH HYBRID REFLECTOR ANTENNAS*

**I N Kartsan[1], P V Zelenkov[2], V N Tyapkin[3], D D Dmitriev[3], A E Goncharov[1]**

[1] Associate Professor, Siberian State Aerospace University (SibSAU), Krasnoyarsk, Russia
[2] Science and Research Department Director, SibSAU, Krasnoyarsk, Russia
[3] Associate Professor, Siberian Federal University (SFU), Krasnoyarsk, Russia

E-mail: kartsan2003@mail.ru

**Abstract.** This paper reviews adaptation algorithms influenced by active interferences in satellite communication systems. A multi-beam antenna is suggested as an adaptive system; it is built on the basis of a hybrid reflector antenna with a 19-element array feed element, which incorporates a modified algorithm for radiation pattern synthesis used for suppressing targeted interferences. As a criterion for this synthesis, antenna gains are used at fixed points. As a result, the size of the objective function and time required for the synthesis can be significantly limited.

**1. Introduction.** Satellite communication systems are becoming increasingly important due to the immensity of service areas and the high availability factor – the ability to provide communication channels in any place on the globe in a short time period. This rapid technological development significantly increases requirements to satellite communication systems. Among the requirements are channel capacity enhancement, channel security, interference protection and flexible traffic. These requirements are especially relevant for the Russian Federation: its territory is considerably extended longitudinally, the population density in many areas is low, there are 11 time zones. The most effective method of implementing these requirements on satellites is the usage of multi-functional adaptive antennas with spatial selection.

However, the usage of these antennas also encounters a number of difficulties. Unlike the simpler classical antenna, the adaptive multi-antenna includes a high number of active expensive elements (controlled phase shifters, attenuators, control loops elements, etc). Consequently, it is necessary to take into consideration not only existing technical characteristics, but economical indicators for selecting an assembly method for these antennas.

Forming multiple beams and implementing spatial selection is possible either on the basis of an adaptive phased array or on the basis of hybrid reflector antennas [1].

* The research was done with the financial support of the Ministry of Education and Science of the Russian Federation in accordance with the agreement from 28.11.2014 № 14.577.21.0155, unique ID project RFMEFI57714X015.

Adaptive phased arrays have significant advantages over other types of antennas in speed maintenance and multitasking [2]. Adaptive phased arrays with digital beamforming enable flexible control over the direction of the main maximums in each of the partial radiation patterns for a wide range of angles; they are capable of forming deep nulls in radiation patterns in the direction of the source of interference. However, adaptive phased arrays weigh more and are larger in size; they are equipped with more complicated hardware – particularly the beamforming arrangement. These factors, consequently, result in significantly higher costs for these systems [3].

The scan angles for adaptive hybrid reflector antennas are smaller but this disadvantage is insignificant for geostationary spacecraft and spacecraft with high elliptical orbits and is fully compensated by the simpler design.

In many situations it is more promising to use an antenna array having a relatively small number of elements as a feed element for focusing the quasioptical element of the reflector. The primary antenna array is commonly known a feed cluster. Hybrid reflector antennas combine the advantages of highly efficient reflector antennas and antenna arrays. Hybrid reflector antennas enable the formation of tunable multipath radiation patterns and radiation patterns with a special shape; they provide the spatial filtering of signals and interferences; and they are capable of adapting to failures of the primary antenna array.

This class of antennas allows combining the advantages of reflector antennas – namely the simplicity of the aperture construction, the possibility of effective transformation and the flexibility typical of adaptive phased arrays. The technical specifications of hybrid reflector antennas are equivalent to those of adaptive phased arrays, the first being significantly less expensive due to the reduction of the number of active elements.

In such antennas the system of adaptation and forming radiation patterns should be conducted in two modes – the multi-beam mode and the contour beam.

In the first case, the coverage of the required service area is provided by a set of narrow beams. The application of a multi-beam radiation pattern enables multiple uses of the frequency resource due to spatial isolation between the beams. In the second case, a radiation pattern which exactly repeats the system's service area is formed. The pattern flexibly redistributes power within a given service area.

Technically, it is possible to arrange these modes either for simultaneous or for individual use. Economically, the combination of these modes is more advantageous since one primary antenna array can be used for both modes.

## 2. Adaptation algorithms for hybrid reflector antennas

The following antenna arrangement can be proposed as a variation of the adaptive hybrid reflector antenna. The antenna array of the feed element is built from 19 feed elements using the cluster principle – each of the 7 beams of the antenna is formed by a set of feed elements (the cluster) (Fig. 1) [4]. Each cluster, consisting of 7 feed elements, is powered in-phase, half of the radiated power is supplied to the central feed element and half is redirected to peripheral feeds.

IOP Conf. Series: Materials Science and Engineering **94** (2015) 012010 doi:10.1088/1757-899X/94/1/012010

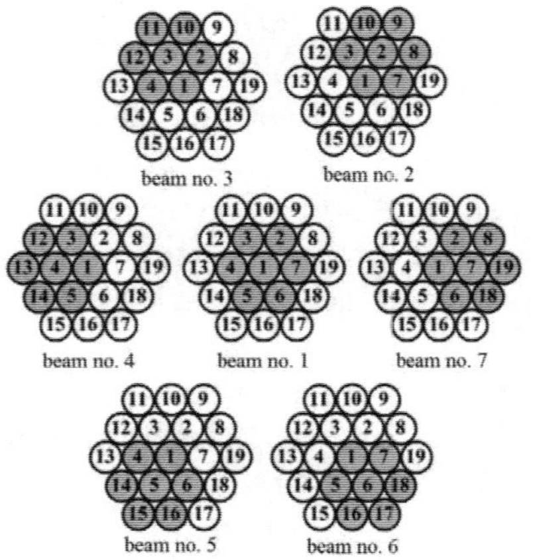


**Figure 1.** Beamforming principal based on cluster scheme of the primary antenna array.

Such a scheme enables beamforming for 7 beams in the radiation pattern for different frequencies, preventing the beams from contacting one another. Nullforming in the radiation pattern of one cluster is possible by powering one of the feed elements in antiphase (Fig. 2) [5]. This method, however, has one significant disadvantage – the null is formed at a fixed direction.

Powering the feed element antiphase results in a significant signal loss: up to 20% from the nominal service area (Fig. 2). Furthermore, to suppress additional interference it is necessary to power an additional feed element in antiphase. This would result in further degradation and the complete loss of antenna characteristics.

Nulling radiation patterns in the direction of the side lobes is possible by increasing the number of feed elements, which form the radiation pattern of a hybrid reflector antenna. However, this method is not always effective, since it may lead to a complication of the antenna's structure and an increase in the weight and the size of the antenna.

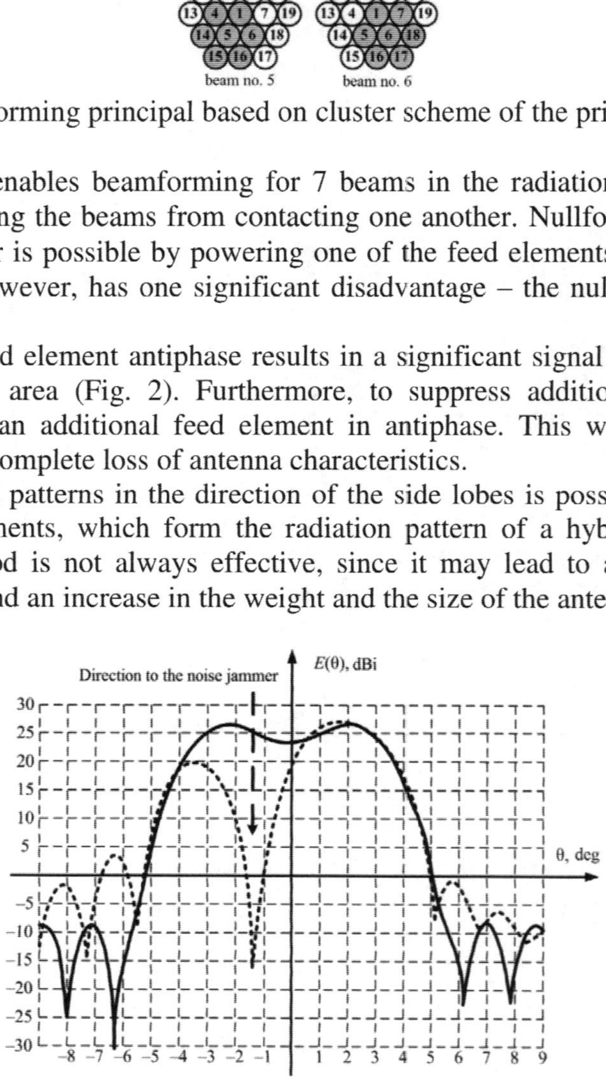


**Figure 2.** Interference suppression from the direction of main lobe in a hybrid reflector antenna by powering one of the cluster members antiphase.

To improve the performance of the antenna, let's synthesize the amplitude-phase distribution of the hybrid reflector antenna's primary array. This will enable to synthesize the desired shape and nulls of the radiation pattern not only in the direction of the main lobe, but in the direction of the side lobes as well. The number of feed elements in the array remains unchanged.

Generally, the radiation pattern in the far field of a multibeam antenna during performance in a cluster can be written as

$$\left|E(W_j,\theta_i,\varphi_i)\right|^2 = \left|\sum_{j=1}^{N_{изл}} E_{ij}\cdot W_j\right|^2, \tag{1}$$

where $W_j = A_j \cdot e^{j\varphi_j}$ is the weighted factor of the feed element. $E_{ij}$ is the radiation pattern from $j$ (feed factor) in $i$ (direction of observation). $\left|E(W_j,\theta_i,\varphi_i)\right|^2$ is the required radiation pattern of the antenna.

It is necessary to limit the amplitude of feed elements depending on the normalization condition of cluster power

$$\sum_{j=1}^{N_{изл}} A_j^2 = 1. \tag{2}$$

$E_{ij}$ is the beforehand known value; it further transforms into the constant for the particular point $\theta_i$, $\varphi_i$ during the synthesis of the adaptive phased arrays. In order to make calculations of the adaptive phased arrays less time consuming, it is possible to limit function (1) by a set of points, which are determined programmatically. A desired value of the signal level for each point is defined.

The set of weighted coefficients can be given as a vector

$$\mathbf{W} = \left[W_1, W_2, ..., W_j\right]^{*T}. \tag{3}$$

The calculation is carried out in a set of points to limit the size of the objective function and the duration of the synthesis of the adaptive phased arrays. This decreases the time needed for calculating the required adaptive phased arrays and improves the efficiency of the whole system.

The directive gain of the antenna in the given points (targets) is taken as a criterion for synthesizing radiation patterns. Additionally, it is necessary to modify the algorithm due to the limited number of points and the need for nullforming in radiation patterns. Let's introduce a concept of the wave channel "target", which will increase or decrease the signal level for a given "target":

w > 0 – the increase of signal level at a given point, tends to the maximum possible value.

w < 0 – the signal suppression at a given point, tends to the lowest possible value.

Based on the aforementioned, we can write the objective function of the algorithm for synthesizing the cluster of adaptive phased arrays as

$$f_i = w_i\left(\left|E(\theta,\varphi)_{цель,i}\right|^2 - \left|E_i(\theta,\varphi)\right|^2\right), \tag{4}$$

where $w_i$ is the weighted factor of the "target". $|E(\theta,\varphi)_{цель,i}|^2$ is the required value of the signal level for a given point in space. $|E_i(\theta,\varphi)|^2$ is the achieved value of the signal level for a given point of space.

Thus, there is a set of values of $f_i$, corresponding to the axis set in a required service area.

The purpose of this algorithm is to minimize the value of function $F(x)$. The parameter value is tried by successive iterations to achieve the minimum value of the objective function.

$$F(\bar{x}) = \max\{ f_1(\bar{x}), f_2(\bar{x}), f_3(\bar{x}), ...., f_k(\bar{x})\}. \quad (5)$$

The equal-amplitude cophasal distribution of antenna arrays (arrays containing 19 feed elements) is taken as the first iteration. Then, the algorithm calculates the absolute gain for each given point on the basis of equal-amplitude distribution, establishing the maximum value of the difference between the required and the actual gain $\Delta E$ = $(|E(\theta,\varphi)_{цель,i}|^2 - |E_i(\theta,\varphi)|^2)$. From a set of ΔE the point at which there is maximum difference in the initial state and the desired ($\Delta E_{\max}$) is found; after this the gradient vector $\left[\frac{\partial E}{\partial W_1}, \frac{\partial E}{\partial W_2}, ...., \frac{\partial E}{\partial W_j}\right]^T = \nabla E$ in the point of maximum difference is determined. Next, the vector parameter **W** is replaced with its new parameter value of $\mathbf{W}_1 = \mathbf{W}_0 \pm \alpha \cdot \nabla E$, where α is the step size defined by software. This process is repeated as long as $\|\mathbf{W}_1 - \mathbf{W}_0\| < \varepsilon$ (ε is given before the beginning of the synthesis process), i.e. until the difference between the desired gain value and the actual gain value for each "target" achieves minimum changes. The number of iterations can be set programmatically, or the algorithm will automatically stops working after a minimum in changes has been attained.

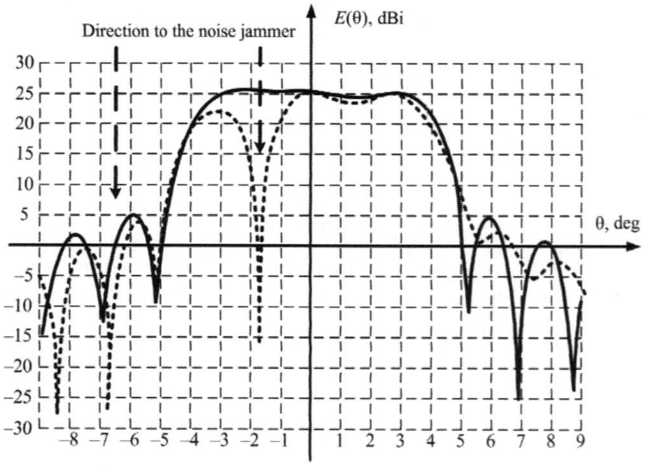


**Figure 3.** Suppressing interference in hybrid reflector antenna from the main and the first side lobes.

Calculation for a cluster of adaptive phased arrays with 7 feed elements is performed using the synthesized algorithm. The calculation was performed for a case of interference in the direction of the main lobe and the direction of the first side lobe (Fig. 3).

IOP Conf. Series: Materials Science and Engineering **94** (2015) 012010 doi:10.1088/1757-899X/94/1/012010

## 3. Conclusion

The following conclusions can be made from the research:

1. Hybrid reflector antennas, equipped with the proposed algorithm for adaptive phased arrays synthesis, have provided nulling in radiation patterns in the direction of the main lobe and in the direction of side lobes of minus 45–50 dB.

2. A decrease in the signal level at10% of the total service area has been recorded.

3. An insignificant difference in the signal level in the main lobe (not exceeding 0.3 dB) has been observed. It was provoked by setting the signal level requirements in the service area which hadn't been affected by the interference.

Thus, the proposed adaptation algorithm for hybrid reflector antennas enables us to fulfill all current requirements for satellite communication systems with minimal expenses.

## References

[1] Ehyaie D 2011 Novel approches to the design of phases array antennas (The University of Michigan) *Dissertation available at:* *http://www.eecs.umich.edu/radlab/html/NEWDISS/Ehyaie.pdf*

[2] Tyapkin V N, Dmitriev D D, Perschin A S 2013 The adaptive algorithm multibeam antenna, built on hybrid-reflector antenna *Journal of Siberian Federal University. Engineering & Technologies* vol.6 **7** pp. 835–844.

[3] Widrow B 1989 Adaptive Signal Processing *Adaptivnaia obrabotka signalov* (Moscow, Russia: Radio i sviaz').

[4] Tyapkin V N, Pershin A S, Dmitriev D D, Moshkina T G 2012 Adaptive multibeam antennas *Vestnik of Siberian State Aerospace University named after academician M.F. Reshetnev* **2(42)** pp. 106–109

[5] Dzhun W I 1989 Adaptive antenna systems with interference nulling on the main lobe of the radiation pattern [Adaptivnye antennye sistemy s podavleniem pomekh po glavnomu lepestku diagrammy napravlennosti] *Foreign radio electronics* [*Zarubezhnaia radioelektronika*] **4** pp. 3–15.

# THE PRACTICAL APPROACH TO THE RELIABILITY ANALYSIS OF THE SOFTWARE ARCHITECTURE OF A COMPLEX COMPANY CONTROL SYSTEM

**I V Kovalev, P V Zelenkov, and S Ognerubov**
Siberian State Aerospace University named after Academician M.F.Reshetnev
31 "KrasnoyarskiyRabochiy" prospect, Krasnoyarsk, 660037, Russia.

E-mail: zelenkov@sibsau.ru

**Abstract.** The practical aspects of the implementation of reliability analysis of the architecture of a complex control system of a company are considered in this article. The comparative analysis for two variants of software architecture using different factors is presented, the relations between the reliability characteristics and the amount of system architecture components and their connections with each other are defined.

## I. Introduction

Software architecture as a research area is wide [1-3]. There are not enough technical and management manuals for the software architecture control for software companies. The wide architecture of a complex company control system (CCCS), which is a member of the group of software systems which operate in real-time and have a very complex (intricate) communication complex (including software versions) is considered as an example in [4, 5]. The following architecture levels are chosen to describe a software system [6]:

processes;
functions;
primitives;
data;
structures;
messages.

To measure the amount of changes which are generated by the subsystem module failures additional data is required. Using that data and the information about the architecture, the system processes and the module changes as a result of a failure it is possible to measure and increase the reliability of a software architecture [7-9].

In the software reliability model [10], one architecture level matches the subsystem's processes and additional architecture levels match the functions, the primitives, the data, the structures and the messages for these processes. Different types of component are placed on different levels and connected using their communication and calling services. The architecture matches the generalized scheme which is presented as an example in [11].

In the given architecture the processes interact through messages using functions and primitives. The functions use the data and the structures. The processes, the functions, the primitives, the data, the structures and the messages are located on different architecture levels in the model. In this architecture, all components of the same type use patterns and the same connection for every component of a given type.

## II. The initial parameters of the architecture levels components

The values of the parameters for the presented model depend on the type of the considered real-time software architecture, the amount of the architecture levels, the components and their relations. Different architectures are considered in order to show how the amount of components and other parameters of a software system influence the reliability. In that case, the amount of functions, primitives and data change that shows the change in the amount of components of an architecture of a software system.

The unconditional and conditional failure probabilities are different for different amounts of components. The unconditional failure probability depends on the amount of code in the components. For example, it can be proportionate to the number of lines of code in the components or can be dependent on the ratio of the amount of changed lines to the amount of all lines in the program. The unconditional failure probability can be estimated as the rate of the component recoveries independent from the recoveries of other components. The conditional failure probability depends on the amount of code in the components and the relations between the components. For example, the conditional failure probability can be higher for a component which depends on a higher amount of other components (under the condition of a static dependence between components) or is connected to a higher amount of other components (under the condition of a dynamic dependence between components). The conditional failure probability can be estimated as the rate of the component recoveries which are generated by other components recoveries.

Two variants of architecture are presented for comparison. Table I and table II show the parameters used in that example [12].

## III. The comparison of architecture variants result analysis

The comparison analysis for two variants of telecommunication software architecture of a complex company control system is done using the dialog system of the architecture estimation [13].

**Table 1.** The values of the parameters of the architecture variant A.

| The architecture levels | The amount of components | The component failure probability | The conditional failure probability on different levels | The access, analysis and recovery time | The percentage of component usage |
|---|---|---|---|---|---|
| The processes | 7 | 0.1 | 0.0 | 5 | 100 |
| The functions | 399 | 0.2 | 0.8 | 7 | 80 |
| The primitives | 138 | 0.05 | 0.4 | 12 | 90 |
| The data | 202 | 0.12 | 0.5 | 12 | 25 |
| The structures | 49 | 0.05 | 0.2 | 11 | 40 |
| The messages | 34 | 0.05 | 0.1 | 14 | 60 |

**Table 2.** The values of the parameters of the architecture variant B.

| The architecture levels | The amount of components | The component failure probability | The conditional failure probability on different levels | The access, analysis and recovery time | The percentage of component usage |
|---|---|---|---|---|---|
| The processes | 7 | 0.1 | 0.0 | 5 | 100 |
| The functions | 215 | 0.35 | 0.9 | 12 | 90 |
| The primitives | 63 | 0.1 | 0.5 | 18 | 110 |
| The data | 278 | 0.1 | 0.4 | 10 | 20 |
| The structures | 49 | 0.05 | 0.2 | 11 | 40 |
| The messages | 34 | 0.05 | 0.1 | 14 | 60 |

The downtime of the system is shown in "Fig." 1. System B has the minimal downtime having a lesser amount of functions and primitives and a larger amount of data. The decrease in the amount of data causes the most downtime for system A. It means that there is a set of the parameter's values in the large telecommunication software architecture, which decrease system's downtime using the permissible architecture changes. That model's characteristic can be used for the development of the most reliable architecture for the large real-time telecommunication software systems [14-17].

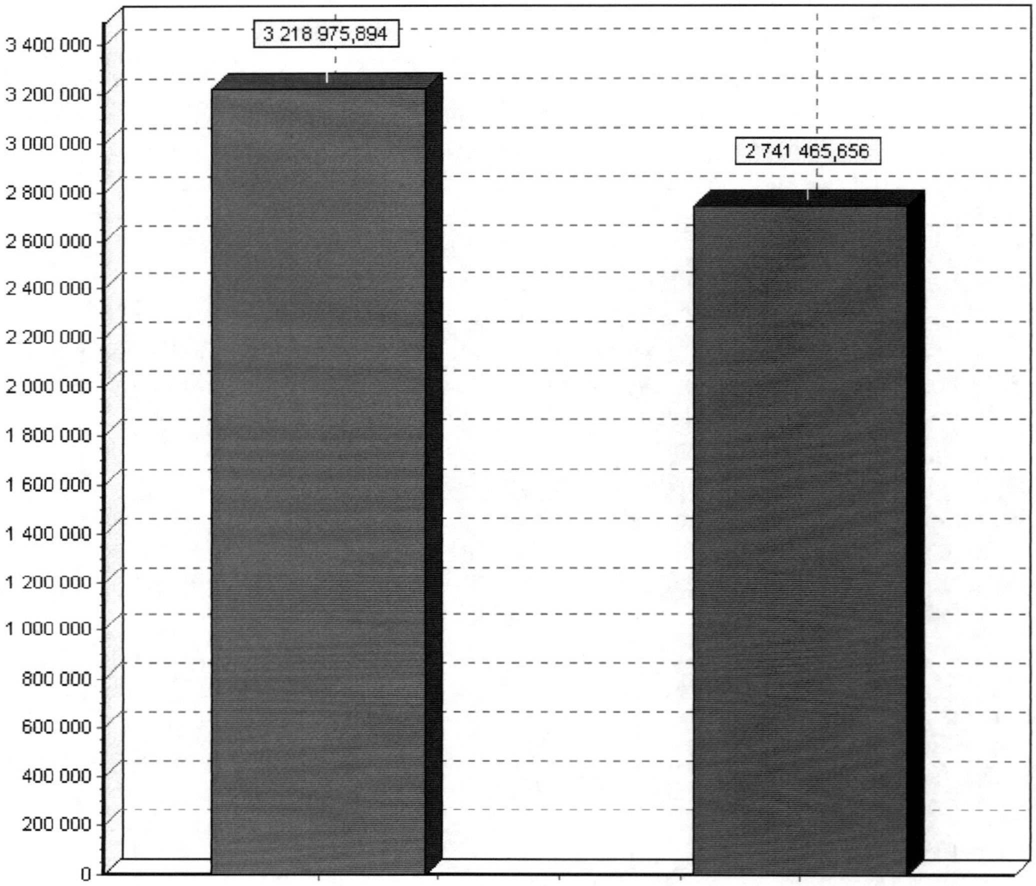


**Figure 1.** The graph of the average systems downtime for both architecture variants.

IOP Conf. Series: Materials Science and Engineering **94** (2015) 012011 doi:10.1088/1757-899X/94/1/012011

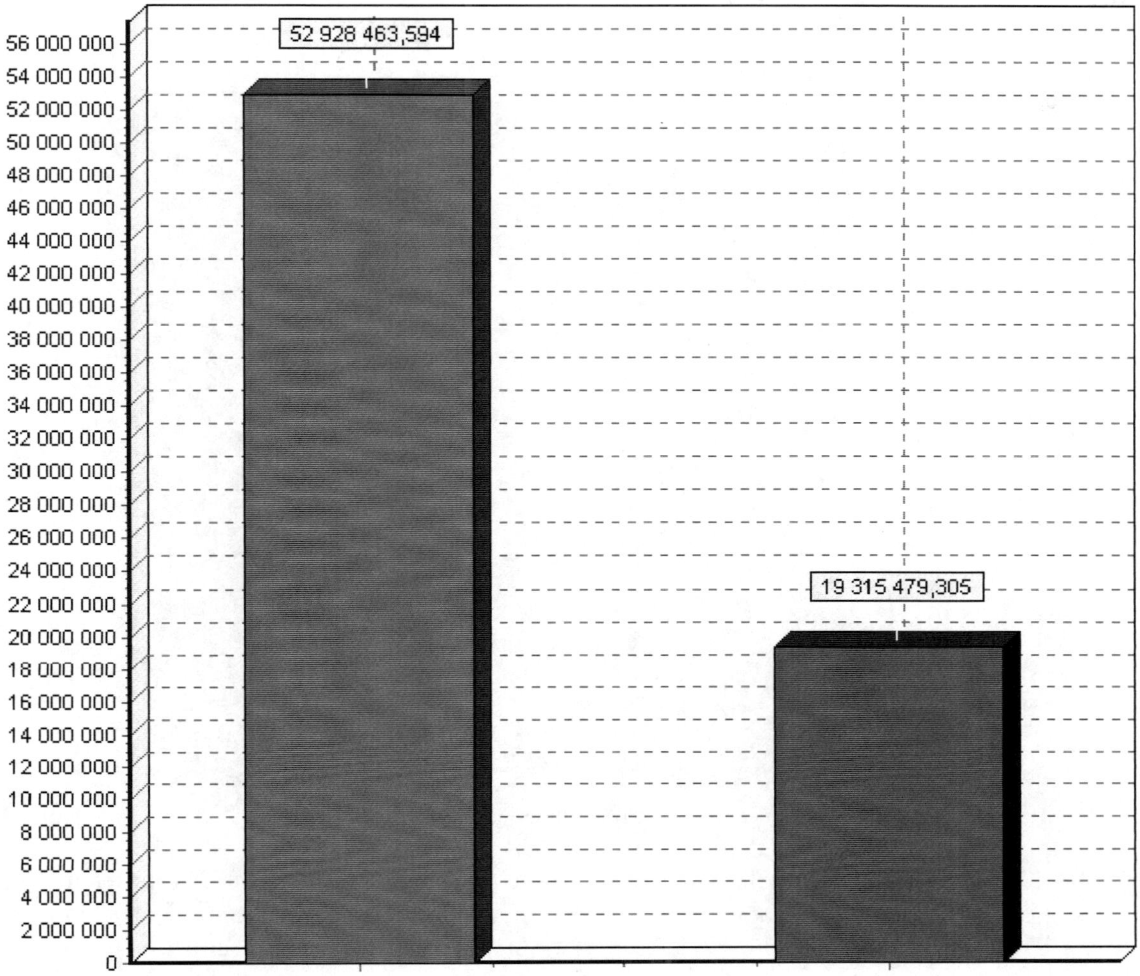


**Figure 2.** The graph of the average system failure appearing time for both architecture variants.

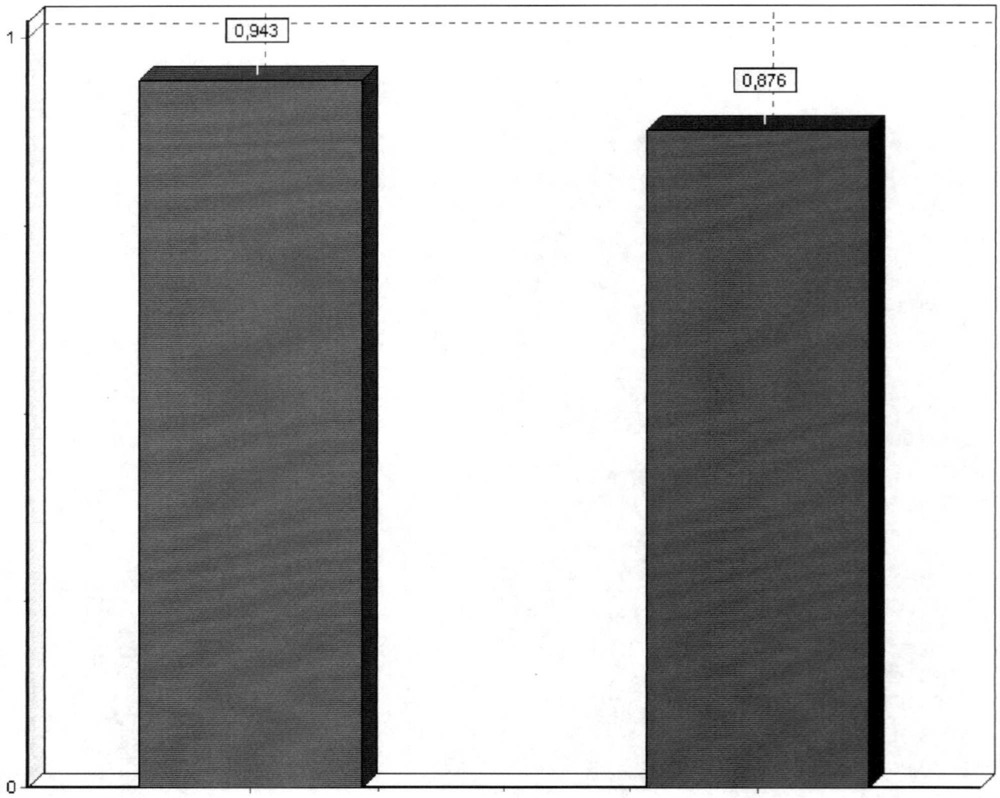


**Figure 3.** The graph of the system readiness coefficient for both architecture variants.

The average system failure appearance time for a software subsystem is shown in "Fig." 2. The average system failure appearance time is increasing then the amount of the components rises. Therefore, architecture variant A consists of 829 components and has the average system failure appearance time of 52 926 463 seconds, while variant B consists of 646 components and has the average system failure appearance time of 315 479 seconds. The average system failure appearing time correlates with the usage time of the components. Therefore a lesser average system failure appearance time does not mean lesser reliability of the large architecture of the real-time data transmission.

The system readiness coefficients for both systems are presented in "Fig." 3. This graph is a generalization of the two graphs described above, because both the average system failure appearance and average systems downtime are used for the system readiness coefficient calculation.

Using the graphs in "Fig." 1 and "Fig." 2 it is hard to decide whether system architecture A or B is more reliable, whereas the graph of the system readiness coefficient for both architecture variants shows that despite higher system downtime, meaning a larger amount of time is required for the system to recover, the architecture variant A is more reliable. The possibility of the fact that in the moment of time t system A will be functioning well is 0.943 and for system B the possibility is 0.876.

It can be seen in "Fig." 4 that the development cost of variant A is higher than variant B (the development cost of a single component is the same for all components, even for components from different architecture levels).

Variant A is the best choice in terms of reliability, but the most important factor is the development cost of a whole system.

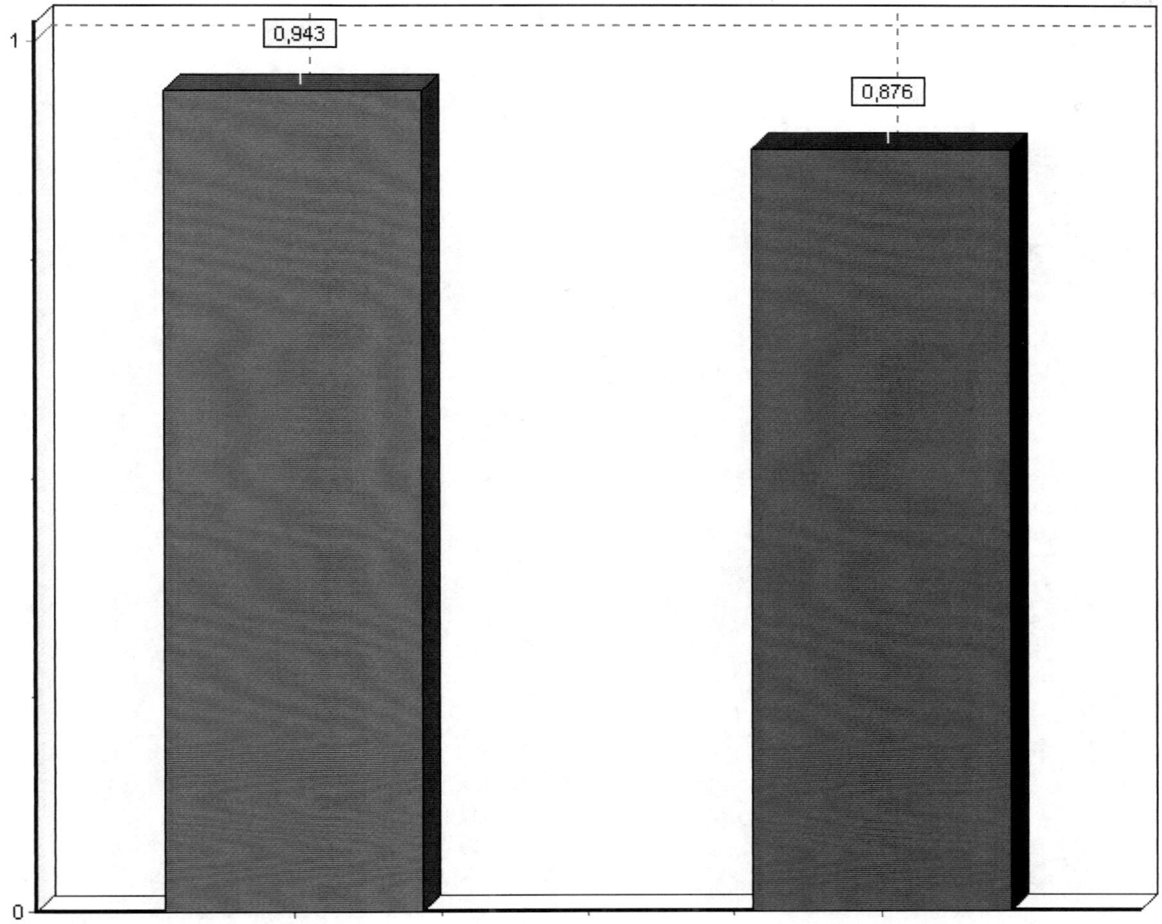


**Figure 4.** The graph of the average system development cost for both architecture variants.

The dialog system of the architecture software reliability express-analysis [10] provides opportunities for more thorough architecture analysis. It can be presented as a graph where the change in the reliability characteristics of an architecture depending on the change of the amount of components on one of many architecture levels is shown. There is a possibility to change the amount of the functions and data for variant A. The dependence of the reliability parameters on the amount of the functions and data graphs are presented in "Fig." 5-7.

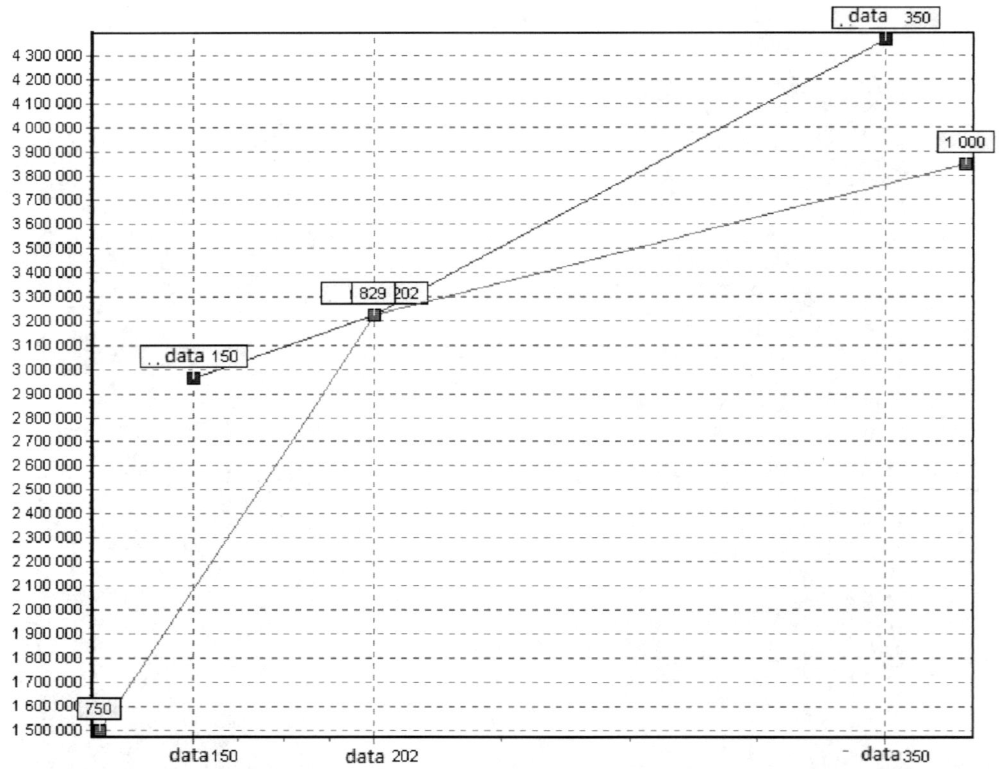


**Figure 5.** The graph of the dependence of the change in the length of system downtime on the change in the amount of the functions and data.

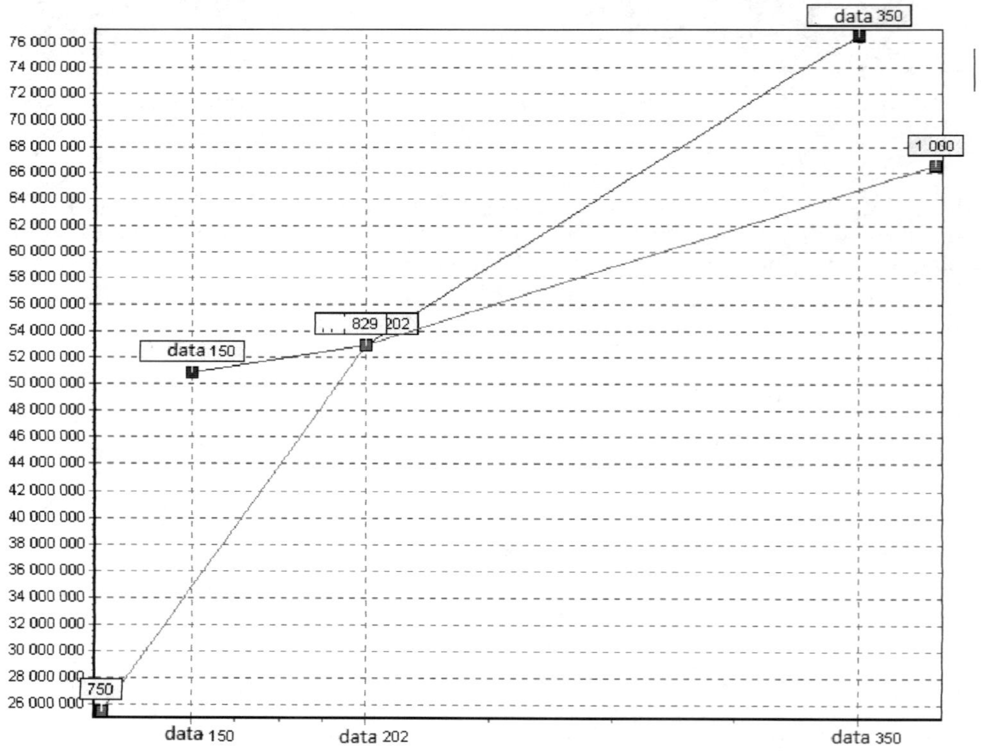


**Figure 6.** The graph of the dependence of the failure appearance time on the change in the amount of the functions and data.

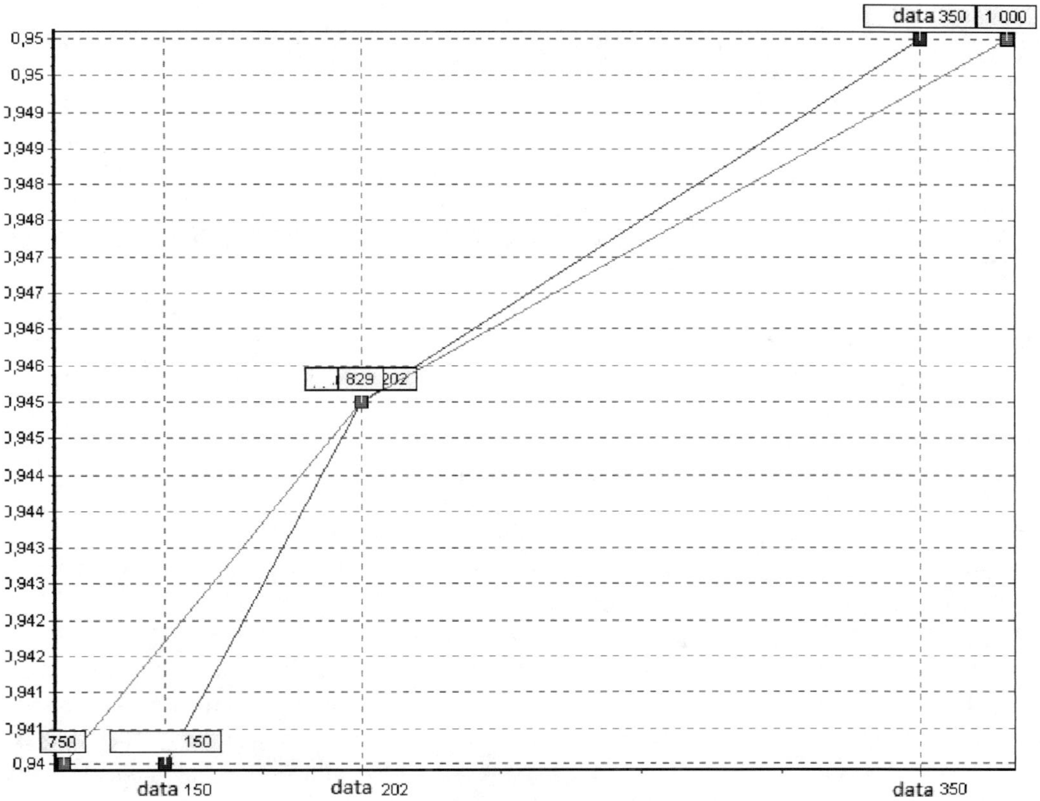


**Figure 7.** The graph of the dependence of the system readiness on the change in the amount of the functions and data.

These graphs show that the length of system downtime, the failure appearance time and the system readiness are rising with the increase in the amount of the components. Using the system readiness graph as a generalized indicator it is possible to make the conclusion that the best results can be achieved by increasing the amount of components on the architectural level "Data". Under the condition of a small decrease in the amount of components, a decrease in the system readiness happens.

The conditional and unconditional probabilities of failure and recovery time as well as the usage time of the components are different depending on the amount of components. The express-analysis system can be used to estimate the software reliability for the possible architectural changes and to choose the most reliable architecture.

The possible selection under the condition of the development of complex software systems and the telecommunication, real-time software created using the client-server architecture is limited by many factors of the environment. These factors include the hardware architecture which cannot be changed, the architecture levels which cannot be divided or united, the code of a component which cannot be changed and many other limitations. Therefore, the subset of the possible changes in the software architecture in real environment usually includes a few variants. The model [10] can be applied to many variants to estimate the software reliability while choosing the best software development course.

## References.

[1] Levendel Y 1990 Reliability analysis of large software systems: Defect data modeling *IEEE Trans. Software Engineering* **16** pp.141-152.

[2] Kovalev I, Zelenkov P, Ognerubov S 2015 The efficiency analysis of automated lines of companies based on DEA Method. *Lecture notes in economics and mathematical systems* **675** pp. 107-115.

[3] Kovalev I, Zelenkov P, Ognerubov S, Brezickaya V, Kovalev D 2014 The practical realization of the software architecture reliability analysis *Proceedings of the Work in Progress Session held in Connection with SEAA 2014 the 40th EUROMICRO Conference on Software Engineering and Advanced Applications and DSD 2014 the 17th EUROMICRO Conference on Digital System Design* (Verona (Italy), August 27-29, 2014) SEA-Publications: SEA-SR-40, pp. 17-18.

[4] Kovalev I V, Zelenkov P V, Karaseva M V, Tsarev M Yu and Tsarev R Yu 2015 Model of the reliability analysis of the distributed computer systems with architecture "client-server" *IOP Conf. Ser.: Mater. Sci. Eng.* **70(1)** art. no. 012009, doi:10.1088/1757-899X/70/1/012009.

[5] Wattanapongsakorn N 2001 Reliability optimization for software systems with multiple applications *FastAbstract ISSRE and Chillarege Corp. Copyright.*

[6] Kovalev I, Younoussov R 2002 Fault tolerant software architecture creation Model based on reliability evaluation *Advances in Modeling & Analysis* vol.7 **3** pp 31-44.

[7] Xie M, Yang B 2000 Regression goodness-of-fit test for software reliability model validation *FastAbstract ISSRE Copyright.*

[8] Lyu M R 1996 Handbook of software reliability engineering *Edited by Michael R. Lyu Published by IEEE Computer Society Press and McGraw-Hill Book Company* 819 p.

[9] Kovalev I, Soustin B, Kardashov V 1988 Multicriteria decision model for modular software structure selection *in proc. of the Intern. AMSE Conf., Lyon, France* pp 1.9-1.11.

[10] Ashrafi N 1992 Optimization models for selection of programs, considering cost & reliability *IEEE Transaction on reliability* vol.41 **2** pp.281-287.

[11] Zahedi F 1991 Software reliability allocation based on structure, utility, price, and cost *IEEE Trans. on Software Engineering* vol.17 **4** pp.345-356.

[12] Kovalev I, Grosspietsch K 2000 Deriving the optimal structure of N-version software under resource requirements and cost *Timing Constraints," in proc. of Euromicro'2000, Maastricht, Netherlands* pp 200-207.

[13] Kovalev I, Vasilenko N, Shabalin N, Davydov I 2000 Using a software reliability model in decision-making support systems for the software structure selection *Advances in Modeling & Analysis* vol.5 **1–2** pp 51-62.

[14] Kovalev I, Popov A, Shipovalov Ju 2000 Optimization models for reliability of telecommunication software systems *Advances in Modeling & Analysis* vol.43 **3–4** pp 41-46.

[15] Rosenberg L et al 1988 Software metrics and reliability *Software reliability engineering was presented at the 9-th International Symposium, "Best Paper" Award.*

[16] Novoy A V, Shtentsel A V 2008 The estimation of reliability of the structure of multi-version software for the processing and control information systems *Vestnik SibGAU* **3(20)** pp. 73–79.

[17] Kovalev I V, Zelenkov P V and Tsarev M Yu 2015 The control of developing a structure of a catastrophe-resistant system of information processing and control *IOP Conf. Ser.: Mater. Sci. Eng.* **70(1)** art. no. 012008, doi:10.1088/1757-899X/70/1/012008.

# THE MATHEMATICAL STATEMENT FOR THE SOLVING OF THE PROBLEM OF N-VERSION SOFTWARE SYSTEM DESIGN*

**I V Kovalev, D I Kovalev, P V Zelenkov, A A Voroshilova**
Siberian State Aerospace University Named after Academician M.F. Reshetnev
31 "KrasnoyarskiyRabochiy" prospect, Krasnoyarsk, 660037, Russia.

E-mail:zelenkov@sibsau.ru

**Abstract.**The N-version programming, as a methodology of the fault-tolerant software systems design, allows successful solving of the mentioned tasks. The use of N-version programming approach turns out to be effective, since the system is constructed out of several parallel executed versions of some software module. Those versions are written to meet the same specification but by different programmers. The problem of developing an optimal structure of N-version software system presents a kind of very complex optimization problem. This causes the use of deterministic optimization methods inappropriate for solving the stated problem. In this view, exploiting heuristic strategies looks more rational. In the field of pseudo-Boolean optimization theory, the so called *method of varied probabilities* (MVP) has been developed to solve problems with a large dimensionality.

## 1. Introduction

Development of high-reliable fault-tolerant systems is an interesting engineering problem having not only technical meaning but also social importance. Systems of this kind determine the stability in social and technical environments, and multiple examples of such systems' crashes prove the strong need for more reliable constructions which can be realized through the use of up-to-date methods and approaches.

The rapid progress of computer technique of late years has made the software an essential part of any complex automated system. The reliability of software component may determine the reliability of whole the hardware-software system. That's why during last years large attention is paid to the development of the methodologies of designing high-reliable software complexes [1-5].

Practically, multi-channel tools increasing the system reliability at the expense of a multiple duplication of certain structure elements are very much in evidence. This approach has given a good account of itself in the designing of hardware parts of complex systems. The use of this methodology leads to a sizable decreasing of appearance probability of random errors having the physical nature. In turn, this approach is not an influence on software reliability, since it doesn't trace so called *dormant* (or sleeping) errors which could arise while writing the program code by a stated specification [6].

* The research was done with the financial support of the Ministry of Education and Science of the Russian Federation in accordance with the agreement № 14.574.21.0041, unique identifier RFMEFI57414X0041

*The multi-version programming*, as a methodology of the fault-tolerant software systems design, allows successful solving of the mentioned tasks. The idea of multi-version programming has been introduced by A. Avizienis in 1977 [7]. The term N-version programming (NVP) used in the literature is of equal meaning and often takes place in papers on the observed methodology. A. Avizienis introduced NVP as *an independent generation* of N≥2 functionally equivalent software modules from the same *initial specification*. The *concurrent execution tools* are provided for such the modules. In cross-check points (cc-points) software modules generate cross-check vectors (cc-vectors). The components of the cc-vectors and the cc-points are to be determined in the specification set.

The use of N-version programming approach turns out to be effective, since the system is constructed out of several parallel executed versions of some software module. Those versions are written to meet the same specification but by different programmers. Where, the writing process of each version of concrete software module in any way must not intersect with or depend on another version code writing. This is done to avoid the presence of same *dormant* (or sleeping) errors in separate software designs. This kind of errors is typical for software components.

The problem of developing the optimal structure of an N-version software system (NVS) is the following: to choose a set of software modules, so as to provide the highest reliability for the system subject to the budget constraint. Since a description of any possible system configuration is made through such the positioning of its components, we can say that an observed problem has the binary essence [8]. Moreover, the existing theory of pseudo-Boolean functions and their optimization contains strong tools for solving problems of this kind [9]. And that fact makes the use of binarization algorithms more affordable.

The process of a problem binarization consists in setting relationships between the system states and the binary space elements. In the case of our system model, we need to determine some Boolean vector the elements of which will characterize the system structure. Each element of such the Boolean vector will signify either presence or absence of corresponding system component [10].

In that way, before starting to describe the exact process of binarization, all the necessary terms should be coined and the presented system model should be overviewed in details.

## 2. Optimization Model for Structuring NVS

The structure of N-version software system is determined consisting of a set of tasks (a set $\mathbf{I}, \mathrm{card}\,\mathbf{I} = I$). All the tasks are divided into classes, i.e. a set of task classes is introduced as well ($\mathbf{J}, \mathrm{card}\,\mathbf{J} = J$).

To solve the tasks belonging to a certain class, there is a software module, which can be realized by any of its versions. Thus, $\mathbf{K}, \mathrm{card}\,\mathbf{K} = J$ – the set of software modules. Let us introduce the vector $\mathbf{S} = \{S_j\}\,(j = \overline{1,J})$, each component of which is equal to a number of module versions ($S_j$ – the number of versions of module solving a task of class *j*) [11].

To describe the task belonging to particular classes, in [12] the authors define sets of tasks for every task class. That is introduced as two-dimensional array in programming terms. Since the numbers of tasks belonging to different classes are not equal, that may cause some difficulties when translating the analytic expressions into a program code.

Here, it is proposed to use only one set the capacity of which is equal to the number of tasks in a system. And each element of this set is equal to the number of class a task belongs to. So, the set **B**, card $\mathbf{B} = I$ is the set of class membership of tasks, i.e. the element $B_i$ of the set **B** presents the number of class the *i*-th task belongs to.

Using the introduced notations, lets us determine a common analytic form of the number of versions solving *i*-th task. If the element $B_i$ of the set **B** is the number of class the *i*-th task belongs to, then an element of the set **S**, the index number of which is equal to $B_i$, determines the number of versions in a module solving *i*-th task. Therefore, this number can be written like this $S_{B_i}$.

Basing on that, we will introduce the Boolean variables $X_s^i$ to describe the control implication of different module versions:

$$X_s^i = \begin{cases} 1, & \text{the } s-\text{th } (s=\overline{1,S_{B_i}}) \text{ version of module } B_i \text{ is used} \\ & \text{to solve the } i-\text{th } (i=\overline{1,I}) \text{ task,} \\ 0, & \text{the } s-\text{th } (s=\overline{1,S_{B_i}}) \text{ version of module } B_i \text{ is } \textbf{not} \\ & \text{used to solve the } i-\text{th } (i=\overline{1,I}) \text{ task.} \end{cases} \tag{1}$$

Expanding the introduced variables into the *implication vector* is the head moment in applying pseudo-Boolean optimization methods to the considered systems design.

Since a vector component number is specified by only one index and we deal with two-index variables, it is necessary to establish an algorithm forming an implication vector and an algorithm determining the component indices of this vector. Following section contains the algorithms to convert a problem of optimal structure design for N-version systems to a problem of pseudo-Boolean optimization and vice versa.

### 2.1. Conversion Algorithms

The algorithm of an implication vector forming acts in the following way (see the scheme on fig. 2). The first component of an implication vector **X** describes the first version of a module to be involved in the solving of the first system task. If the software module which solves the first task has more than one version, the next component of vector **X** characterizes the second version of the first task module. In this way, all the versions of all software modules are overhauled.

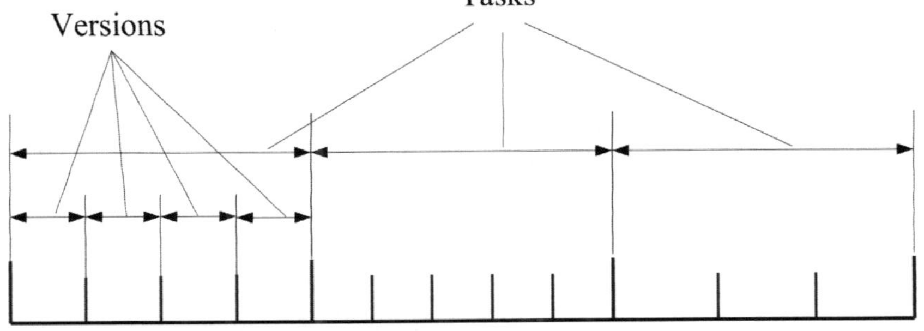


**Figure 1.** An example of the implication vector.

Hence, in order to determine the number of a vector **X** component, being aware of corresponding number of a task $i$ and a number of version $s$, it is necessary to sum the number of versions in the modules solving the first ($i$-1) tasks and to add $s$ to obtained value.

Analytically this conversion appears as follows:

$$pos = \begin{cases} s & , if\ i = 1, \\ \sum_{j=1}^{i-1} S_{B_j} + s & , if\ i > 1. \end{cases} \tag{2}$$

In order not to recalculate the first sum (the number of versions in the first ($i$-1) tasks) every time when optimizing a system, it would be better to count those sums depending on different $i$ and to memorize them in an index vector:

$$G_i = \sum_{j=1}^{i} S_{B_j}$$ , or in recurrent form

$$G_i = \begin{cases} S_{B_1} & , if\ i = 1, \\ G_{i-1} + S_{B_i} & , if\ i > 1. \end{cases} \tag{3}$$

Therefore, the value of the $i$-th component of vector **G** equals the number of versions in modules solving the tasks from the first up to the $i$-th. It results from this that the value of the last vector **G** component is equal to $n$ – the implication vector dimensionality, i.e.

$$\sum_{i=1}^{N} S_{B_i} = n .$$

Once the index vector is introduced, the analytic record of a calculation of the implication vector component number takes the form of the following:

$$pos = \begin{cases} s & , if\ i = 1, \\ G_{i-1} + s & , if\ i > 1. \end{cases} \tag{4}$$

The reverse conversion task (a conversion of the implication vector component number to the numbers of task and version) consists of the consecutive determining of $i$ and $s$. The flowchart of this algorithm is show on the figure 2.

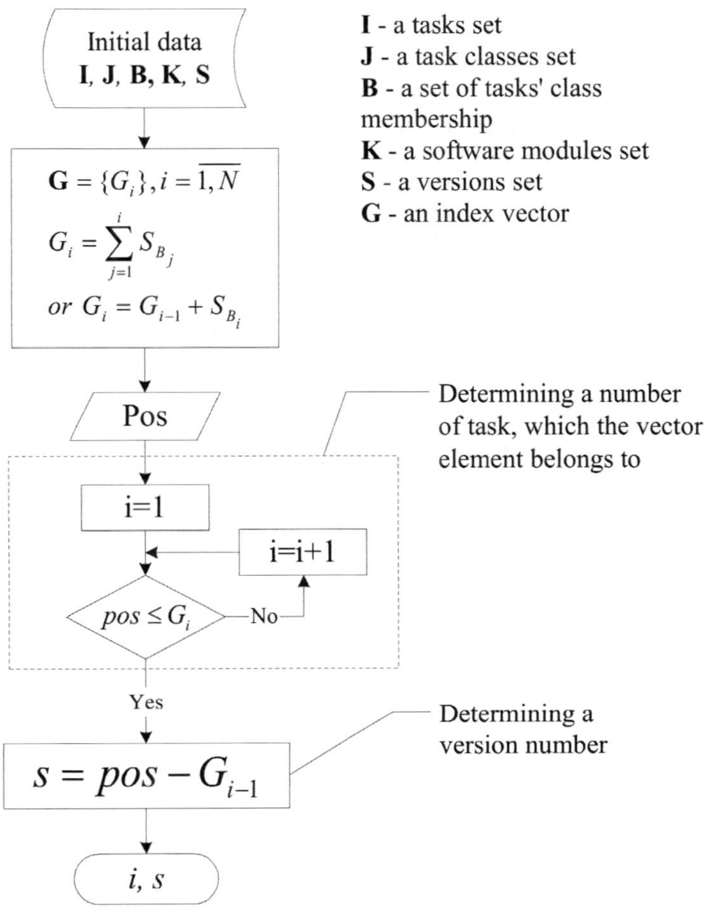


**Figure 2.** A conversion of the implication vector component number into the numbers of task and version.

Since the $i$-th element of the index vector equals the number of versions in modules solving the tasks from the first up to the $i$-th, the task number is determined by comparing the index of the implication vector component with the elements of the index vector. The comparison is being made from the first element of the index vector till the last one sequentially. And when the value of the parameter *pos* turns out to be less than or equal to the value of the index vector component, the required task number takes the value of this element number.

Then subtracting the number of versions in all the tasks from the first up to the ($i$-1)-th (equal to $G_{i-1}$) from *pos* we get a version number of software module solving the $i$-th task corresponding to *pos*.

The two presented algorithms [12] are the core of applying binary approach to solve the stated problem. Thus, having received the tools for a problem conversion, it became possible to use the methods developed within the confines of pseudo-Boolean optimization study. Some the features of the considered problem are discussed in the following section. Basing on this the conclusions about the relevant methods is made.

### 2.2. The mathematical Statement of the Problem

The converting algorithms considered above allow to describe the NVS structure in a form of a Boolean vector. As it was noticed previously, the optimal design of control system is held subject to different parameters: the reliability (it should be as big as possible), the cost (it should be as small as possible or at least it shouldn't exceed some limit), the allocation & scheduling and so on [14].

In terms of optimization theory, a system reliability function of a system structure is nothing else but an objective function. And conditions imposed on the system structure are the constraints set to limit the objective function domain [13]. Since we are able to associate a system structure with a Boolean vector, an objective function is a pseudo-Boolean one. And an optimization problem becomes a pseudo-Boolean one too.

In the framework of the presented model we will use a system reliability function as the objective function and the system cost will be the constraint imposed on the system [15-19]. In analytic form this problem can be written as follows:

$$\max R = \prod_{i=1}^{I} R_i ,$$

where

$$R_i = 1 - \prod_{s=1}^{S_{B_i}} (1 - R_{B_i s})^{X_s^i}$$

subject to

$$\sum_{i=1}^{I} \sum_{s=1}^{S_{B_i}} X_s^i \cdot C_{B_i s} \le B .$$

Here, $R_{B_i s}$ and $C_{B_i s}$ are the reliability and the cost of the software version $s$ from module which solves the task of class $B_i$

### 3. Conclusion

The problem of structuring an N-version software system is specified by the binary character, what made it plausible to apply the methods of pseudo-Boolean optimization. Within the limits of the discrete optimization a set of methods and algorithms has been proposed. Thus, having received the tools for a problem conversion, it became possible to use the methods developed within the confines of pseudo-Boolean optimization study.

### References

[1] Laprie J-C et al 1987 Hardware- and Software-fault tolerance: definition and analysis of architectural solutions *Proceedings of the IEEEp* pp. 116-121.

[2] Anderson T, Barrett P A, Halliwell D N, Moudling M L 1985 An evaluation of software fault tolerance in a practical system *Proc. Fault Tolerant Computing Symposium* pp. 140-145.

[3] Scheer S, Maier T 1997 Towards dependable software requirement specifications *In: Daniel, P. (ed.) Proceedings of SAFECOMP* (New York)

[4] Kovalev I 1994 Optimization-based design of software of the spacecraft control systems *In: "Modelling, Measurement and Control* (AMSE Press), *B* vol.56 **1** pp. 29-34.

[5] Kovalev I V, Engel E A, Tsarev R Ju 2007 Programmatic support of the analysis of cluster structures of failure-resistant information systems *Automatic Documentation and Mathematical Linguistics* vol.41 **3** pp. 89.

[6] Keene S J 1994 Comparing hardware and software reliability *Reliability Review* **14(4)** pp. 5-21.

[7] Avizienis A 1995 The methdology of N-version programming *In: Software fault tolerance* (edited by M.R. Lyu, Wiley) pp. 23-47.

[8] Antamoshkin A, Schwefel H P, Torn A, Yin G and Zilinskas A 1993 System analysis, design and optimization. *Ofset Press* (Krasnoyarsk) 312 p.

[9] Antamoshkin A N et al 2013 Random search algorithm for the p-median problem *Informatica* **3(37)** pp. 127–140.

[10] Kazakovtsev L, Stanimirovic P, Osinga I, Gudima M and Antamoshkin A 2014 Algorithms for location problems based on angular distances. Advances in Operations Research. Articale ID 701267. 12 pages. - http://www.hindawi.com/journals/aor/raa/701267/

[11] Kovalev I 1998 Optimization problems when realizing the spacecrafts control *In: Advances in Modeling and Analysis* vol. 52 **1-2** pp. 62-70.

[12] Kovalev I V, Dgioeva N N, Slobodin M Ju 2004 The mathematical system model for the problem of multi-version software design *Proceedings of Modelling and Simulation, MS'2004 AMSE International Conference on Modelling and Simulation*, (MS'2004. AMSE, French Research Council, CNRS, Rhone-Alpes Region, Hospitals of Lyon. Lyon-Villeurbanne).

[13] Kovalev I, Grosspietsch K-E 2000 Deriving the optimal structure of N-version software under resource requirements and cost *Timing Constraints* (Proc. Euromicro' 2000, Maastricht, 2000, IEEE CS Press) pp. 200-207.

[14] Kovalev I et al 2013 The minimization of inter-module interface for the achievement of reliability of multi-version software *Proceedings of the 2013 International Conference on Systems, Control and Informatics* (SCI 2013), Venice, Italy, September 28-30 pp. 186-188.

[15] Ashrafi N et al 1994 Optimal design of large software-systems using N-version programming *IEEE Trans. on Reliability* vol.43 **2** pp. 344-350.

[16] Ashrafi N, Berman O 1992 Optimization models for selection of programs, considering cost and reliability *IEEE Trans. on Reliability* vol.41 **2** pp. 281-287.

[17] Ashrafi N, Berman O 1993 Optimization models for reliability of modular software systems *IEEE Trans. on Software Engineering* vol.19 **11** pp. 1119-1123.

[18] Kovalev I 1995 Optimal base software composition of the spacecrafts control system *In: "Advances in Modelling and Analysis, C"* (AMSE Periodicals) vol.47 **3** pp. 17-26.

[19] Kovalev I V et al 2002 Fault-tolerant software architecture creation model based on reliability evaluation *Advanced in Modeling & Analysis* (Journal of AMSE Periodicals) vol.48 **3-4** pp. 31-43.

# THE PSEUDO-BOOLEAN OPTIMIZATION APPROACH TO FORM THE N-VERSION SOFTWARE STRUCTURE*

**I V Kovalev, D I Kovalev, P V Zelenkov, A A Voroshilova**
Siberian State Aerospace University named after Academician M.F. Reshetnev
31 "KrasnoyarskiyRabochiy" prospect, Krasnoyarsk, 660037, Russia.

E-mail:zelenkov@sibsau.ru

**Abstract.** The problem of developing an optimal structure of N-version software system presents a kind of very complex optimization problem. This causes the use of deterministic optimization methods inappropriate for solving the stated problem. In this view, exploiting heuristic strategies looks more rational. In the field of pseudo-Boolean optimization theory, the so called method of varied probabilities (MVP) has been developed to solve problems with a large dimensionality. Some additional modifications of MVP have been made to solve the problem of N-version systems design. Those algorithms take into account the discovered specific features of the objective function. The practical experiments have shown the advantage of using these algorithm modifications because of reducing a search space.

## 1. Introduction

The use of N-version programming approach turns out to be effective, since the system is constructed out of several parallel executed versions of some software module [1,2]. Those versions are written to meet the same specification but by different programmers. Where, the writing process of each version of concrete software module in any way must not intersect with or depend on another version code writing.

The problem of developing the optimal structure of an N-version software system (NVS) is the following [3]: to choose a set of software modules, so as to provide the highest reliability for the system subject to the budget constraint. Since a description of any possible system configuration is made through such the positioning of its components [4,5], we can say that an observed problem has the binary essence [6]. Moreover, the existing theory of pseudo-Boolean functions and their optimization contains strong tools for solving problems of this kind [7]. And that fact makes the use of binarization algorithms more affordable.

The process of a problem binarization consists in setting relationships between the system states and the binary space elements. In the case of our system model, we need to determine some Boolean vector the elements of which will characterize the system structure. Each element of such the Boolean vector will signify either presence or absence of corresponding system component [8].

To derive an optimal dependability solution by means of an systematic, the exhaustive comparison algorithm would mean that all potential system configurations have to be tentatively generated, checked for the fulfillment of the side conditions and processed to

*The research was done with the financial support of the Ministry of Education and Science of the Russian Federation in accordance with the agreement № 14.574.21.0041, unique identifier RFMEFI57414X0041

compute the corresponding overall system reliability. This usually would cause a computing complexity that is untractable even for the most modern high speed computers: if, e.g., we consider a system consisting 64 modules, all of which are to be triplicated, thereby selecting each of the module versions from 5 different candidate modules, we would have to consider $[5!/3!\cdot 2!]^{64} = 10^{64}$ different system configurations! Assuming e.g. 1 nsec for processing each system configuration (of course, a value by far too optimistic!), the resulting $10^{55}$ sec of needed computation time would exceed the estimated age of the universe of about $10^{17}$ sec by many orders of magnitude! Therefore, here only stochastic search methods appear possible to provide, in a heuristic way, an optimal solution [9-12].

## 2. The Method of Varying Probabilities

In the field of pseudo-Boolean optimization theory, the so called *method of varied probabilities* has been developed to solve complicated problems, especially those ones with a large dimensionality [6-8]. The method of variable probabilities (MVP) presents a family of heuristic algorithms based on the common scheme: in order to find an extremal solution of a pseudo-Boolean optimization problem, a *probability vector* of dimensionality of sought solution vector is formed. Each component of the probability vector presents a probability of assigning a *one* value to the correspondent component of a Boolean vector. In the terms of developing NVS structure, it looks like a probability to include a version-candidate into the system structure.

The initial values of the probability vector components describe a situation when every software version has the equal probability to be included into the system structure. Then, at a computational phase, random decisions are generated according to the probability distribution specified by means of the probability vector. Each time the objective function is calculated in several random points, the values of the probability vector components are updated, so changing a probability distribution form. The way of changing these values defines a separate algorithms of MVP scheme. The common approach for updating a probability vector can be characterized by the rule: the better result received with a one-valued binary vector component the bigger probability is assigned for it to get the value of one in the final solution.

This scheme can be augmented whether by some special methods for updating the probability vector or through involving the peculiar procedures of generating random solutions at a computational phase of an algorithm. This paper discusses the two methods for updating the probability vector (ARSA and Modified ARSA ver. 1) and the two procedures of generating random solutions (the independent generation of random points and the generation of non-zero solutions) giving thus as a result four different realizations (algorithms) of MVP.

The Adaptive Random Search Algorithm (ARSA) plays a role of the background for the rest of the algorithms of MVP scheme [10]. Initially, ARSA has been developed for the problem of pattern recognition to select an informative subsystem of attributes. The main disadvantage of this algorithm is a potential problem of updating values of probability vector components. Namely, in some cases it is possible to get the values of intermediate solutions which do not let the probability vector components to be changed. To correct the defect, the modification of ARSA has been developed (Modified ARSA ver.1). The statistical data of applying the modified version of ARSA display the better behavior of the algorithm when solving problems of developing a structure of NVS.

Next, applying to the stated optimization problem, ARSA doesn't provide a technique of avoiding zero-solutions when solving the problem of designing NVS structure. To protect an algorithm against spending both computational and time resources for calculating the objective function values in the points of this kind, the particular technique of generating random non-zero solutions has been developed. This technique is utilized in the MVP based algorithm named NVS MVP (mentioning the strict field of using the algorithm).

Making use of both of the mentioned enhancements gave a great raise in the efficiency of applying the MVP based algorithms to the problem of NVS structure development. The statistical results presented in the final part of the paper show it. Different algorithms have been tasted on the same optimization problem with the same quantity of objective function calls.

The objective function of the presented optimization problem has several specific features which can assist to reduce a search domain, thus allowing to decrease the searching time. The objective function as a function of the whole system reliability represents the product of reliabilities of separate software modules. Consequently, when a reliability of any of the modules is equal to zero the overall system reliability turns into zero value also. Physically, it represents a case when there are no versions chosen for (at least) some of the software modules. The implication vector components corresponding to such the software modules will be assigned zeroes as well. Obviously, it is necessary to avoid computing the objective function in such the points.

The number of system structures having at least one software module without versions-candidates assigned can be determined as the difference of the number of all the possible structures and the quantity of the structures which provide every software module with at least one candidate, i.e. $N_0 = N_{all} - N_{R>0}$. The number of all possible structures is determined through the dimensionality of an implication vector $n$ as follows $N_{all} = 2^n$. The second intermediate value is found basing on the multiplication principle from combinatorics as a number of all possible structures with software module combinations each without one of them (that with no versions assigned). Formally, it is described in the following way: $N_{R>0} = \prod_{i=1}^{I}(2^{k_i} - 1)$, where $I$ is the number of software modules, $k_i$ represents a number of versions for the $i$-th software module.

Then, the final expression determining a sought value looks like this:

$$N_0 = 2^n - \prod_{i=1}^{I}(2^{k_i} - 1).$$

The value of this expression depends on an overall number of candidates (a dimensionality of the optimization problem), a number of software modules $I$ and the numbers of versions for each of the software module ($k_i$, $i = \overline{1,I}$). In general case, this expression takes grand values counting up to $0.9N_{all}$, i.e. 90% of all the possible solutions. This means in this case that in order to find a solution it is sufficient to search through only 10% of the definitional domain of the objective function.

Unfortunately, there is the other side of the question making this result not so optimistic. Namely, for the problems of large dimensionalities reducing the search domain to 10% means diminishing the dimensionality of a problem by very small value. For instance, for a test problem of dimensionality $n$=117, avoiding all the null-valued points lowered the problem dimensionality only down to $n_{R>0} = 116$.

Nevertheless, exploiting this feature of the objective function has given satisfactory results when applying the algorithms of the method of varied probabilities (MVP). The modification of the MVP with the ability of avoiding null-valued points called NVS MVP has its own way of generating random points at iterational steps of the algorithms. In NVS MPV, random points are generated so that to provide each software module with at least one version.

At every iterational step, the whole implication vector generated is concerned as consisting of parts each describing the structure of a separate software module. Thus indeed, random vector generating consists of generating of random structures of modules. This approach allows having only non-zero solutions in result.

## 3. The Random Search of Boundary Points

Another stochastic algorithm to optimize the structure of NVS is the algorithm of *random search of boundary points* [8]. It is based on the proved fact that a solution of the stated optimization problem is a so called *boundary point*. Or in terms of binary space topology, a point neighboring to the set of infeasible solutions. Such a point describes a system structure which can not be updated through including a software versions additionally without violating the resource conditions, i.e. no version can not be added to a system structure of this kind paying attention to restrictions. The algorithm of random search of boundary points constitutes a generating of multiple boundary solutions and comparing the objective functions values in them.

The constraint in this optimization problem partitions the whole binary space into two domains – the domain of solution satisfying the constraint function and a set of points not satisfying to the constraint. It is shown that these domains represent the connected sets and that a solution of correspondent optimization problem is a point neighboring to the set of infeasible solutions. This kind of solution can be called a *boundary point*.

Basing on the results stated above, it is clear that it is sufficient to search among only boundary points in order to find the best value of the objective function. Thus, the problem of finding a best solution becomes a problem of an exhaustive search on the boundary points set.

The following is the algorithm of generating boundary point for the problem of developing the optimal structure of NVS (Fig. 1).

Different boundary points can be reached using this algorithm when different combinations of ways to choose $i$ at the second step of the algorithm will be followed.

Hence, the algorithm of searching boundary points will have the following scheme.

1. The initializing step: $i$=0.
2. Determine a boundary point $\mathbf{X}_{bi}$ (b – as an index means “boundary”).
3. Calculate the objective function value $F_i=F(\mathbf{X}_{bi})$.
4. If the stopping condition is satisfied go to p. 5, otherwise $i=i+1$ and go to p. 2.
5. The solution is $F^* = \max_i F_i$.

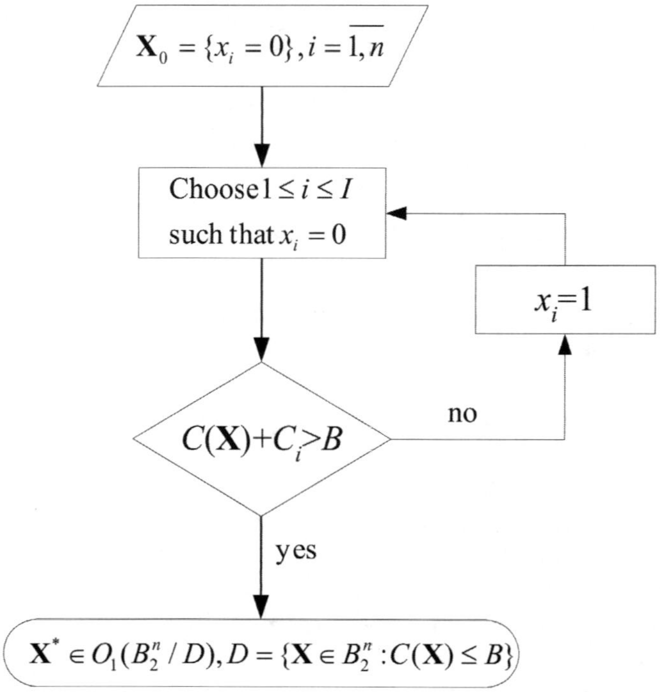


**Figure 1.** Generating a boundary point.

Separate variations of the *algorithm of boundary points search* may differ from each other in a stopping condition and in ways of reaching boundary when generating boundary points. For the optimization problems of high complexity it is more rational to use stochastic version of the algorithm when boundary points are reached in a random way and this process is executed repeatedly.

## 4. The Comparison of the Random Optimization Procedures

Concluding the paper, let us cite the comparative data of the computational results for different random optimization procedures. To gather such the information, the presented algorithms have been used to solve the test NVS structure optimization problems [9-12]. The efficiency of the random search algorithms has been judged by the values of the objective and constraining functions.

The problem of dimensionality 117 has been chosen as the test problem, i.e. the developed software system included 117 software versions. It's worth mentioning that every of the random search algorithms needed approximately same period of time for calculating under equal conditions. That's why the time has not been set as an efficiency characteristic.

Table 1 contains the computational results of algorithms testing [13,14]. The best searching capabilities have been revealed with the use of NVS MVP algorithm and the algorithm of boundary points search. The latter displayed the highest stability of the solutions found, although using NVS MVP it is sometimes possible to find more reliable system configurations.

**Table 1.** The results of random search algorithms working.

| | Budget constraint B | Number of iteraions | The random search algorithms | | | |
|---|---|---|---|---|---|---|
| | | | NVS MVP | | Random search of boundary points | |
| | | | $R(X^*)$ | $C(X^*)$ | $R(X^*)$ | $C(X^*)$ |
| 1. | 00 | 15000 | *0.7872* | 789 | *0.8074* | 796 |
| | | | *0.7916* | 791 | *0.7998* | 797 |
| | | | *0.7907* | 791 | *0.8177* | 794 |
| | | 30000 | *0.8136* | 786 | *0.8318* | 794 |
| | | | *0.8054* | 775 | *0.8331* | 797 |
| | | | *0.8118* | 784 | *0.8377* | 798 |
| 2. | 00 | 15000 | *0.9040* | 850 | *0.9149* | 899 |
| | | | *0.9207* | 896 | *0.9164* | 899 |
| | | | *0.9039* | 887 | *0.9148* | 898 |
| | | 30000 | *0.9076* | 867 | *0.9192* | 897 |
| | | | *0.9082* | 875 | *0.9167* | 897 |
| | | | *0.9155* | 890 | *0.9177* | 892 |
| 3. | 000 | 15000 | *0.9701* | 995 | *0.9622* | 993 |
| | | | *0.9523* | 986 | *0.9609* | 998 |
| | | | *0.9546* | 989 | *0.9635* | 998 |
| | | 30000 | *0.9651* | 994 | *0.9652* | 997 |
| | | | *0.9554* | 988 | *0.9661* | 995 |
| | | | *0.9712* | 997 | *0.9631* | 996 |

## 5. Conclusion

The problem of structuring an N-version software system is specified by the binary character, what made it plausible to apply the methods of pseudo-Boolean optimization. Within the limits of the discrete optimization a set of methods and algorithms has been proposed. The search capabilities of each of the algorithms realized have been investigated by solving the test problems. It was shown that the modification of the method of varying probabilities for NVS MVP together with the algorithm of boundary points search provide the best searching capabilities concerning the time efficiency and the solution quality.

## References

[1] Laprie J-C et al 1987 Hardware- and Software-fault tolerance: definition and analysis of architectural solutions *Proceedings of the IEEE* pp. 116-121.

[2] Scheer S, Maier T 1997 Towards Dependable Software Requirement Specifications *In: Daniel, P. (ed.) Proceedings of SAFECOMP, New York (1997)*

[3] Kovalev I 1994 Optimization-based design of software of the spacecraft control systems *In: "Modelling, Measurement and Control, B"* vol. 56 №1 pp. 29-34.

[4] Kovalev I V, Engel E A, Tsarev R Ju 2007 Programmatic support of the analysis of cluster structures of failure-resistant information systems *Automatic Documentation and Mathematical Linguistics* vol.41**3** pp. 89.

[5] Avizienis A 1995 The methdology of N-version programming *In: Software fault tolerance* (edited by M.R. Lyu, Wiley) pp. 23-47.

[6] Antamoshkin A, Schwefel H P, Torn A, Yin G and Zilinskas A 1993 System Analysis, Design and Optimization. *Ofset Press, Krasnoyarsk* 312 p.

[7] Antamoshkin A N et al 2013 Random Search Algorithm for the p-Median Problem *Informatica* **3(37)** pp. 127–140.

[8] Kazakovtsev L, Stanimirovic P, Osinga I, Gudima M and Antamoshkin A 2014 Algorithms for location problems based on angular distances *Advances in Operations Research* vol. 2014. Articale ID 701267. 12 pages.
- http://www.hindawi.com/journals/aor/raa/701267/

[9] Kovalev I. 1998 Optimization problems when realizing the spacecrafts control *In: Advances in Modeling and Analysis, C* vol. 52 **1-2** pp. 62-70.

[10] Kovalev I V, Dgioeva N N, Slobodin M Ju 2004 The mathematical system model for the problem of multi-version software design. *Proceedings of Modelling and Simulation, MS'2004 AMSE International Conference on Modelling and Simulation* (MS'2004. AMSE, French Research Council, CNRS, Rhone-Alpes Region, Hospitals of Lyon. Lyon-Villeurbanne).

[11] Kovalev I, Grosspietsch K-E 2000 Deriving the Optimal Structure of N-version Software under Resource Requirements and Cost *Timing Constraints* (Proc. Euromicro' 2000, Maastricht, 2000, IEEE CS Press) pp. 200-207.

[12] Kovalev I et al 2013 The Minimization of Inter-Module Interface for the Achievement of Reliability of Multi-Version Software *Proceedings of the 2013 International Conference on Systems, Control and Informatics (SCI 2013)* (Venice, Italy, September 28-30 2013) pp. 186-188.

[13] Kovalev I 1995 Optimal base software composition of the spacecrafts control system *In: "Advances in Modeling and Analysis, C"* (AMSE Periodicals) vol. 47 **3** pp. 17-26.

[14] Kovalev I V et al 2002 Fault-tolerant software architecture creation model based on reliability evaluation *Advanced in Modeling & Analysis* (Journal of AMSE Periodicals) vol. 48 **3-4** pp. 31-43.

# SCIENTIFIC AND EDUCATIONAL CENTER "SPACE SYSTEMS AND TECHNOLOGY"

**I V Kovalev, Y Y Loginov, P V Zelenkov**
Siberian State Aerospace University named after Academician M.F. Reshetnev
31 "Krasnoyarskiy Rabochiy" prospect, Krasnoyarsk, 660037, Russia.

E-mail: loginov@sibsau.ru

**Abstract.** The issues of engineers training in the aerospace university on the base of Scientific and Educational Center "Space Systems and Technology" are discussed. In order to improve the quality of education in the Siberian State Aerospace University the research work of students, as well as the practice-oriented training of engineers are introduced in the educational process. It was made possible as a result of joint efforts of university with research institutes of the Russian Academy of Science and industrial enterprises. The university experience in this area promotes the development of a new methods and forms of educational activities, including the project-oriented learning technologies, identifying promising areas of specialization and training of highly skilled engineers for aerospace industry and other institutions. It also allows you to coordinate the work of departments and other units of the university to provide the educational process in workshops and departments of the industrial enterprises in accordance with the needs of the target training. Within the framework of scientific and education center the students perform researches, diploma works and master's theses; the postgraduates are trained in advanced scientific and technical areas of enterprise development.

The Siberian State Aerospace University (SibSAU) for many years trains engineers for the aerospace industry. Here are the main competitive advantages of the University:

- Unique location of the university in the Siberian center of satellites production and developed relations of the University with the space enterprises;
- Presence of all levels of vocational education including secondary, higher, additional and postgraduate;
- Wide range of educational programs that meet the main objectives of the space industry development;
- Integrated system of aerospace education, based on a combination of theoretical study and production work on the base enterprises or research work in academic institutions, have proven very effective and adapted to modern educational standards and market conditions;
- Wide range of the scientific fields of fundamental and applied research on the creation of advanced space systems; the presence of leading scientific schools in the main areas of research and development;
- Presence in SibSAU required production and test facilities for the development and manufacture of spacecraft.

The University conducts applied research in the design and development of advanced platforms unmanned spacecrafts, small satellites; spacecraft control systems in orbit and methods of use of space-based information systems.

The Scientific and Educational Center "Space Systems and Technology" (SEC) developed in cooperation with the JSC "Information Satellite Systems". The center implements an innovative training program of engineers for JSC "ISS" on the basis of project-oriented learning technologies. JSC "ISS" is one of the leading enterprises of the Russian space industry and has a technology of complete cycle of space systems creation from design to control automatic spacecraft in all orbits - from low circular to geostationary.

The Space Systems and Technologies SEC has been engaged in the design and assembly of a series of scientific and educational small satellites (SmallSat) for a number of years. The project is carried out by a team of students, postgraduate students, young scientists, and specialists of JSC ISS; the team is also responsible for conducting scientific and technological experiments in space. The SEC includes the Student Design Bureau for the design of small satellites, "clean room" for the assembly and testing of small satellites, laboratory of prototyping and satellite electronic systems, laboratory of mechatronic systems and precision mechanics, which equipped with modern high-precision control, measuring and test equipment. The SEC's equipment enables the assembly and testing of satellite mechanical systems, electronic equipment prototyping, conducting vacuum and climate tests, researching spacecraft electronic equipment.

The development of program on creation a series of technological, scientific and educational small spacecrafts provides to undergraduate and graduate students an unique opportunity to take part in the development, design and assembly of satellites in the development of new engineering solutions, creating elements of space systems and space experiments. In the field of development and improvement of spacecraft's elements and nodes the University actively participates in the development and testing of advanced integrated onboard systems control, improvement of on-board power supply system of spacecrafts and developing large-scale transformable satellite structures [1,2].

Here are the objectives and tasks of the program for building a series of scientific, educational, and technological microsatellites:

- Development of an integrated system of engineering education (distance learning systems, laboratory sessions, using special control ground stations designed for the microsatellites, etc.);
- Implementation of a design-oriented educational technology for the preparation of aerospace engineers; forming their professional competencies (student participation in the design and assembly of satellites, service systems, and scientific instruments);
- Scientific experimentation in space (space methods for monitoring the Earth's natural resources, multifunctional nanomaterials, exploitation of high-temperature superconductors and other smart materials in space);
- Technological development: obtaining flight qualification for advanced service systems, instruments and satellite elements (attitude determination and control subsystem, electric power subsystem, thermal control subsystem, and other subsystems with an increased lifetime).

Currently, on the Earth's orbit are two student satellites "Yubileyniy" and "MiR" (Fig.1). All organizations that participated in the development and assembly of the small satellite acquired significant technological experience for future contribution to the training of engineers for the national space industry.

The monitoring and operational control of the small satellites is performed at the students' Satellite Control Center (SCC), which is located at the university. Students obtain telemetry data from the satellite in a real-time mode, learning to decode it and control the satellite. During their course, engineering students have a unique opportunity to directly control, receive, and process telemetry from the following Russian small satellites – YUBILEYNIY, MOZHAETS, CHIBIS, BAUMANETS-2 and satellites of the Technical University of Berlin – DLRSAT, TUBSAT. The Control Center also provides access to the Express-AM program simulator, imitating the flight of a modern functioning telecommunication satellite. The simulator completely imitates the operation of all service subsystems of EXPRESS-AM and is used for conducting tests on the onboard control system, learning how to operate a real satellite. This is a joint development of SibSAU and JSC ISS. The students' SCC is an essential ground segment for the orbital group of technological, research, and educational small satellites, assembled by the university and its partners.

**Figure 1.** Assembling student's small satellite "MiR".

Besides accomplishing existing scientific-technical and experimental tasks, the small satellites are designed to perform an educational function, which is just as significant as the scientific investigation. The production process of these satellites is essential for the preparation of aerospace engineering specialists, who undergo training through a project-oriented educational technology course. This is the first time such a course is introduced at a Russian aerospace university. It has been mentioned that students of SibSAU take part in all stages of satellite production, from writing design documentation to controlling the satellite's orbital path. Implementing the project allows to identify the students who have shown their best effort and enjoyed laboratory work not only during their main curricula, but in related extracurricular activities. This experience shows that such students subsequently become the most qualified and responsible professionals [3].

The construction project of the students' satellite is divided into twelve directions. This models the number of main satellite subsystems – the thermal control subsystem, the electric power subsystem, the attitude determination and control subsystem, etc. The scientific work of each member of the student team must be done in one of these areas. When the student accomplishes his first year on the project, he or she becomes a tutor for younger student of the same direction, sharing the acquired experience. The final results of the student's scientific work in the selected direction, including the graduation project and diploma thesis, is the design and manufacturing of a functional satellite component or constructional element. This part will be installed inside an actual satellite, which will be then launched into space. One of the major projects in 2014, which was attended by students, has been associated with the production of precision structural elements of telecommunications satellites with high-modulus composite materials geometrically stable in space.

A specialized internet-portal has been developed at *www.smka.sibsau.ru* along with a projects management system for project monitoring. This enables the remote coordination of joint projects between different higher education facilities, bringing together students from different cities. The project-and-team education technology enables the student to acquire a high degree of preparation, guaranteeing the alumni will have significant competitive advantages on the labor market as effective workforce at space engineering and high-technology manufacturing enterprises. Typically, after finishing the course most of the graduates are offered employment in the field of their specialization.

In the process realization of scientific and educational projects becomes important the inter-university cooperation with other aerospace universities. Each of the participants, solving their specific problems, contributes to the common fund their experience and their achievements. The inter-university partnership has great potential for expansion of tasks and enhances the level of student training of different specialties. As a result of inter-university collaboration the tasks in the organization of promising scientific and educational spacecraft and satellite systems are solved. The joint development and manufacture of the satellite, its service systems, special and scientific equipment are discussed. Takes place the joint satellite control, the preparation and use of satellite data in the educational and research processes. The block diagram of the ground control of the educational satellite group is shown in Figure 2.

IOP Conf. Series: Materials Science and Engineering **94** (2015) 012014 doi:10.1088/1757-899X/94/1/012014

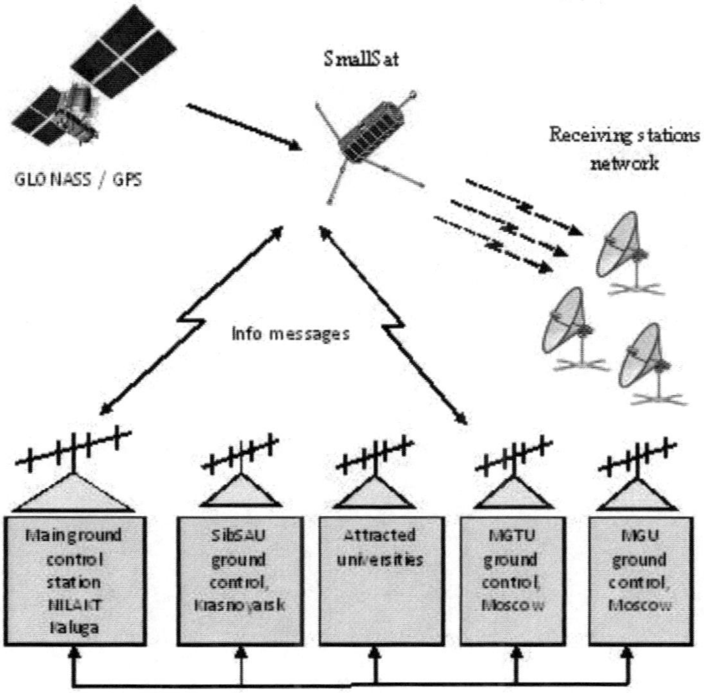


**Figure 2.** Block diagram of the ground control system of research and educational satellites

The experience of creating and using Russian scientific and educational satellites shows great potential and importance of the inter-university partnerships development in these activities. Significant potential of university research combined with creative energy, innovative thinking of students and a good command of modern computer technologies can give new impetus to the development of the national space exploration. Immediate participation of students in the practical work on the design, construction, testing, preparation for the launch and flight control the research and educational satellites significantly increases the level of engineers training to work in the aerospace industry and reduces the risks of care of young professionals in other fields of activity. The practical experience gained by students in the process of work on the creation of university satellites significantly reduces the time to adapt them to perform specific industrial problems on the space industry and the formation of the young specialist as an authoritative member of the labor collectives.

## Conclusion

Thus, the university established a successful integrated system of training engineers for the space industry on the basis of scientific and educational center using the unique technologies of project-and-team students' work, balancing between innovations and traditions, education and scientific research and maintaining the flexibility in educational trajectory construction. The use of learning technologies based on the scientific and educational centers allows improving the quality of engineers training for aerospace industry and the efficiency of joint research work on the development of new techniques and technologies through the development of scientific, educational and industrial potential of participants.

IOP Conf. Series: Materials Science and Engineering **94** (2015) 012014 doi:10.1088/1757-899X/94/1/012014

## References

[1] Kovalev I and Loginov Y 2011 *Higher Education in Russia*, **6** 16
[2] Kovalev I, Loginov Y, Zelenkov P 2015 *IOP Conf. Series: Materials Science and Engineering* **70** 012010, doi:10.1088/1757-899X/70/1/012010
[3] Kovalev I, Loginov Y, Zelenkov P 2015 *Proced. Soc. and Behav. Sci.* **174** 537, doi:10.1016/j.sbspro.2015.01.701

# FAST DETERMINISTIC ALGORITHM FOR EEE COMPONENTS CLASSIFICATION*

**L A Kazakovtsev[1], A N Antamoshkin[1,2] and I S Masich[1]**
[1] Department of System Analysis and Operation Research, Siberian State Aerospace University named after academician M.F. Reshetnev, Krasnoyarsk, Russia
[2] Department of Economics and Information Technology of Management, Siberian Federal University, Krasnoyarsk, Russia

E-mail: levk@bk.ru

**Abstract.** Authors consider the problem of automatic classification of the electronic, electrical and electromechanical (EEE) components based on results of the test control. Electronic components of the same type used in a high-quality unit must be produced as a single production batch from a single batch of the raw materials. Data of the test control are used for splitting a shipped lot of the components into several classes representing the production batches. Methods such as k-means++ clustering or evolutionary algorithms combine local search and random search heuristics. The proposed fast algorithm returns a unique result for each data set. The result is comparatively precise. If the data processing is performed by the customer of the EEE components, this feature of the algorithm allows easy checking of the results by a producer or supplier.

## Introduction

Supplying the electronic units of the complex technical systems with EEE components of the proper quality is one of the most important problems for increasing the whole system reliability. Moreover, for reaching the highest reliability of an electronic unit, the EEE components of the same type must have equal characteristics which assure their coherent operation. The highest homogeneity of the characteristics is reached if the EEE components are produced as a single production batch from a single batch of the raw materials. The critically important units are integrated from EEE components manufactured as a special production lots with special quality requirements [1, 2].

The characteristics of a lot of the components are checked via destructive and nondestructive tests [1, 3]. Data of such tests can be used for analyzing the lot homogeneity [3]. For splitting the EEE components into several assumed production batches, the k-means method can be used [4, 5, 6].

American and European EEE component manufacturers produce components of special quality classes, Military and Space [7, 8]. Russian manufacturers do not form a special class of components for use in space systems [1, 2].

The k-means problem can be classified as a continuous problem of the location theory [9, 10, 11]. The aim is to find $k$ points (centers, centroids, medoids) in a $d$-dimensional space

* Results were obtained in the framework of the state task № 346 of the Ministry of Education and Science of the Russian Federation

such that the total distance from each of the data vectors (known points, measurement result vectors) to the nearest of $k$ chosen centers reaches its minimum:

$$F(X_1,...,X_k)=\sum_{i-1}^{N}\min_{x\in\{x_1,...,x_k\}}\|A_i - X\|_2^2. \quad (1)$$

Quality requirements of the EEE components in space systems are so high that the the range of characteristics measured via quality check tests is very narrow and the quality class and assumed production batch of each component in the lot must be determined via analysis of difference (distance) of test result data vectors which slightly exceed the precision of the measurement tools. Thus, results of each measurement form a finite (discrete) set of possible values defined by the measurement tool precision.

The squared Euclidean norm is most popular distance metric used for calculating differences (distances) in a normalized space of characteristics [9]. Using the rectilinear (Manhattan) [12] norm as a distance metric in the k-means model allows to reach results of the same precision as the precision of data vectors. In this case, the value of each coordinate of the result coincides with the value of the same coordinate of one of the data vectors [9, 13]. Moreover, the result of the k-means problem rectilinear metric are more stable under the influence of the outliers in the data which exist due to measurement errors and defective components in the lot. Other approach which allows to achieve the results of the same precision as data vectors is solving the *k*-medoid problem [14, 15]. In this problem, the cluster centers which minimize the total distance are searched among the data vectors only.

The k-means method uses the ALA procedure (Alternating Location-Allocation) which includes two simple steps:

**Algorithm 1.** ALA procedure.

Required: data vectors $A_1...A_N$, $k$ initial cluster centers $X_1...X_k$.

1. For each center $X_i$, determine its cluster $C_i$ as a subset of the data vectors for which this center $X_i$ is the closest one.

2. For each cluster $C_i$, recalculate its center $X_i$.

$$X_i = \arg\min_{X} \sum_{y\in C_i}\|x\text{-}y\|.$$

3. Repeat Step 1 unless Steps 1, 2 made no change in data.

Traditional usage of the k-means method with squared Euclidean metric ($l_2$) has one important advantage: in this case, calculating a center of a cluster is a simple problem solved in one iteration as calculating the mean value for each coordinate of all data vectors in the cluster. These mean values are coordinates of the new cluster center [9]. If the *i*th cluster center $X_i=(x_{i,1},...,x_{i,d})$ is a vector in a $d$-dimensional space and data vectors $A_j=(a_{i,1},...,a_{j,d})$, $j=\overline{1,N}$ also have $d$ dimensions then the new cluster center is calculated as [9]:

$$x'_{i,k}=\sum_{y\in C_i} y_k / |C_i|, k=\overline{1,d}.$$

In the ALA procedure with the rectilinear metric, each coordinate of the cluster center is calculated as the median value of this coordinate among all data vectors which belong to the cluster. This process can be described as follows.

**Algorithm 2.** Calculating *i*th cluster center (median) in case of the $l_1$ metric.

1. For each $k=\overline{1,d}$ do:

1.1. Sort vectors $A_i = (a_{j,1},\dots a_{j,d}) \in C_i$ in acceding order of the $k$th coordinate, form a sequence $a'_{1,k},\dots,a'_{|C_i|,k}$. Here, $|C_i|$ is the power of the set (cluster).

1.2. Calculate $m = \left[ |C_i| / 2 \right]$. Store $x'_{i,k} = a'_{m,k}$. Here, $[\bullet]$ is the integer part.

1.3. Next iteration 1.

2. Return $X'_i = (x'_{i,1},\dots,x'_{i,d})$

This algorithm returns a value of new center which has values of each coordinate coinciding with a coordinate of some data vector.

In the case of the *k*-medoid problem, procedure of determining of each cluster center requires the exhaustive search among all data vectors in the cluster. However, many researchers propose faster analogous local search procedures [17, 18, 19] which do not guarantee an exact solution.

Except special cases, the k-means and k-medoids problems are *NP* hard and require global search [20].

The result of the ALA procedure depends on the choice of the initial cluster centers. Known k-means++ algorithm [21] has an advantage in comparison with the chaotic choice of the initial centers and guarantees the statistical preciseness of the result $O(\log(p))$. However, such preciseness is insufficient for many practically important problems. For such cases, researchers propose various recombination techniques for initial center sets [9].

The ALA procedure can be optimized with use of many techniques. For example, sampling procedures [22] solve the k-means problem for the randomly selected subset of the data vectors and use the achieved result as an initial set of centers for solving the original problem. Authors propose various algorithms for streaming data processing [4] applicable for big data analysis and many other methods.

The dependence of the results of the ALA procedure on the initial centers seeding is a serious problem for the reproducibility of the classification algorithm results: depending on the initial centers seeding, different algorithm starts classify the same data vectors as elements of various clusters. For the EEE component production batches classification problem, this means that various EEE component belong to the same or different production batches depending on the initial seeding. Thus, an algorithm for solving k-means problem which returns a stable result is preferred.

The Information Bottleneck Clustering method (IBC) is a deterministic method for solving the cluster analysis and classification problems able to achieve perfect results in many cases.[23]. This algorithm starts from considering each data vector as a separate cluster. Then, clusters are removed one-by-one until the desired clusters quantity remains. Each time, the algorithm eliminates such cluster that its elimination gives the smallest increment of the objective function value. For the k-means problem, this algorithm eliminates the cluster center which gives the smallest total distance from data vectors to the closest remaining centers. Such algorithms are extremely slow [23]. The genetic algorithms with greedy heuristic initially designed for the discrete k-median problem on a network [24] are compromise variants. However, they are not determined algorithms. In [25], author proposes an approach for adaptation of these algorithms for the continuous location problems. The idea of such approach can be described as follows [10].

**Algorithm 3.** Genetic algorithm with greedy heuristic for k-median problems.

Required: data vectors $A_1 \dots A_N$, population size $N_p$.

1. Form (randomly or using the k-means++ procedure) $N_p$ various initial solutions $\chi_1,...\chi_{N_p} \subset \{\overline{1,N}\}$, $|\chi_i|=k\forall i=\overline{1,N_p}$. Each of such solutions is a set of $k$ data vectors used as the initial solutions of the ALA procedure. Calculate the fitness function value (total distance) $F_{fitness}(\chi)$ for each of the initial solutions using Algorithm 3 and store the fitness function values to $f_1,...,f_{N_p}$.

2. If the stop conditions are satisfied then STOP. Return solution $\chi_{i^*}$, with the smallest value $f_i$. For the final solution, th ALA procedure (Algorithm 1) runs.

3. Choose randomly two indexes $k_1, k_2 \in \{\overline{1,N}\}, k_1 \neq k_2$.

4. Form an interim solution $\chi_c=\chi_{k_1} \cup \chi_{k_2}$.

5. If $|\chi_c|>k$ then go to Step 7:

6. Calculate $j^*=arg \min_{j\in\chi_c} F_{fitness}(\chi_c \setminus \{j\})$. Exclude $j^*$ from $\chi_c$ : $\chi_c=\chi_c \setminus \{j^*\}$. Go to Step 5.

7. If $\exists i \in \{\overline{1,N_p}\}: \chi_i=\chi_c$ then go to Step 2.

8. Choose index $k_3 \in \{\overline{1,N_p}\}$. First, two indexes $k_4, k_5 \in \{\overline{1,N}\}$ are chosen randomly, then if $f_{k_4}>f_{k_5}$ then $k_3=k_4$ else $k_3=k_5$.

9. Replace $\chi_{k_3}$ and the corresponding fitness function value: $\chi_{k_3}=\chi_c$, $f_{k_3}=F_{fitness}(\chi_c)$. Go to Step 2.

Steps 5-6 of this algorithm realize the greedy heuristic, the successive elimination of the centers from an interim solution.

An analogous heuristic was proposed by Kuehn and Hamburger in 1963 [16]. The IBC method is based on the same principle of the successive elimination of the clusters from an interim solution [23]. Both method of Kuehn and Hamburger and the IBC chose an unfeasible solution coinciding with the whole data vectors set as the initial solution.

The fitness function for the k-means problem can be calculated for an initial or interim solution as follows:

**Algorithm 4.** Calculating fitness function value $F_{fitness}(\chi)$.

Required: solution $\chi$.

1. Run Algorithm 1 with initial centers set $\{A_i | i \in \chi\}$ resulting with centers set $\{X_1,...,X_p\}$.

2. Return $F_{fitness}(\chi)=\sum_{i=1}^{N} w_i \min_{j\in\{1,k\}} L(X_j, A_i)$.

This algorithm is a computationally intensive procedure. Other approach [25] is based on the immediate usage of the total distance from data vectors to the closest center in the interim solution as the fitness function value for Algorithm 3 . In this case, Step 1 of Algorithm 4 is omitted. In fact, this approach solves a $k$-medoid problem and adjusts the solution with the ALA procedure at the final iteration of the greedy heuristic. Such approach is much faster. However, it reduces the preciseness.

An advantage of the IBC method is its determinancy. This method does not use any random values and each start of the algorithm results in the same set of cluster centers. The quality checks of the EEE components lots is a process which involves two parts, a manufacturer or supplier and a customer or a specialized testing center. The calculation performed by the customer must be reproducible. The exact reproduction of the results is impossible when using algorithms which include random search elements. The IBC method slows down the calculation.

In [11], authors propose a modification of the greedy heuristic used in Step 6 of Algorithm 3. This method uses points in the *d*-dimensional space instead of data vector indexes as an alphabet of the genetic algorithm. Such points represent the cluster centers. Authors entitle this modification Algorithm with Floating Point Alphabet. Results of the genetic algorithm with this heuristic are much more precise than results of the modification described in [25] (in [25], authors propose solving a k-medoid problem). At the same time, the iterations in Algorithm 3 with this heuristic are performed much faster than Algorithm 4 as the greedy heuristic. Moreover, decrease of the computational complexity of Step 6 of Algorithm 3 makes solving large-scale problems possible. In [11], authors report results of solving problems with up to 560000 data vectors. Such heuristic can be described as follows.

**Algorithm 5.** Greedy heuristic for the GA with floating point alphabet (used instead of Steps 4-7 of Algorithm 3).

Required: set of data vectors $V=(A_1,...,A_N)\in \mathrm{R}^d$, number $k$ of clusters, two "parent" center sets $\chi_{k_1}$ and $\chi_{k_2}$, parameter $\alpha$.

1. Form interim solution $\chi_c=\chi_{k_1}\cup\chi_{k_2}$. Run ALA algorithm from initial solution $\chi_c$. Store its result to $\chi_c$.

2. If $|\chi_c|=k$ then start ALA procedure from initial solution $\chi_c$, STOP and return $\chi_c$.

2.1. Calculate distances to the closest element of $\chi_c$:

$$d_i=\min_{X\in\chi_c} L(X,A_i)\,\forall i=\overline{1,N}.$$

Form clusters around each center in $\chi_c$. $C_i=\arg\min_{X\in\chi_c} L(X,A_i)\,\forall i=\overline{1,N}.$

Calculate distances from each data vector to the second closest center in $\chi_c$.

$$d_i=\min_{Y\in\chi_c\setminus\{C_i\}} L(Y,A_i)\,\forall i=\overline{1,N}.$$

3. For each center $X\in\chi_c$, calculate $\delta_X=F(\chi_c\setminus\{X\})=\sum_{i:X\in C_i}(D_i-d_i)$.

4.1. Calculate $n_\delta=\max\{[\alpha(|\chi_c|-k)],1\}$. Sort values $\delta_X$ in acceding order and choose a subset $\chi_{e\lim}=\{X_1,...,X_{n_\delta}\}$ of $n_\delta$ data vectors with minimal values $\delta_X$.

4.2. For each $j\in\{\overline{2,|\chi_{e\lim}|}\}$ do: if $\exists q\in\{\overline{1,(j-1)}\}$ : $L(X_j,X_q)<L_{min}$ then eliminate $X_j$ from set $\chi_{e\lim}$. Here, $L_{\min}=\min_{X\in\chi_c}\{\max\{L(X,X_j),L(X,X_q)\}\}$.

4.3. Set $\chi_c=\chi_c\setminus\chi_{e\lim}$.

4.4. Reallocate data vectors to the closest centers: $C_i^*=\arg\min_{X\in\chi_c} L(X,A_i)\,\forall i=\overline{1,N}.$

4.5. For each $X\in\chi_c$ if $\exists i\in\{\overline{1,N}\}$: $C_i=X$ and $C_i^*\neq X$ then recalculate center $X^*$ of cluster $C_X^{clust}=\{A_i\,|\,C_i^*=X,i=\overline{1,N}\}$. Store $\chi_c=(\chi_c\setminus\{X^*\})\cup\{X\}$.

5. Go to Step 2.

Value $\alpha$ is an important parameter of this heuristic. This value determines the percentage of the superfluous cluster centers eliminated in a single iteration. Authors propose value 0.2. Bigger values make the algorithm run faster and reduce its preciseness. Small values $\alpha$ make it work as Algorithm 4 and eliminate a single center at each iteration. We used $\alpha=0.25$.

This heuristic combines the greedy heuristic [24, 25, 3] with elements of the modified ALA procedure performed at each iteration and allows to eliminate up to 20-25% superfluous cluster centers until the required quantity of centers remains. Algorithm 4 requires performing $p(k_0\text{-}k)$ runs of the ALA procedure (here, $k_0$ is the initial centers quantity). Algorithm 5 reduces quantity of iterations down to $O(\log(k_0\text{-}k))$. Moreover, each iteration does not require performint the whole ALA procedure. Instead, its separate optimized elements (location – allocation) are performed.

It can be easily seen that if $k_0=N$ then both Algorithm 5 and IBC method start their initial iterations analogously: number of cluster centers coincides with the number of data vectors. Moreover, if $k_0=N$ then choosing th initial centers is not random: all data vectors are chosen as the initial centers. Thus, the following deterministic algorithm can be proposed.

**Algorithm 6.** New deterministic greedy heuristic algorithm.

Required: set of data vectors $V=(A_1,...,A_N)\in \mathrm{R}^d$, number $k$ of clusters, parameter $\alpha$.

1. Set $\chi_c = V$.

2. Run Algorithm 5 starting from Step 2.

Such algorithm was successfully applied for middle-scale problems, up to $N$=6500 data vectors. The results are gathered in Table 1. We used example problems from the UCI library [26] and problems with real data of EEE components examination. For small-scale problems, the results are shown in comparison with the IBC method, genetic algorithm with greedy heuristic and genetic algorithm with recombination of fixed length subsets. We used various distance metrics. In addition, k-medoid problems [14, 15] were solved. For several problems, the results of new algorithm are insignificantly worse than the results of the IBC. At the ame time, time needed for problem solution reduces many times. In Table 1, all results of new algorithm are shown for $\alpha = 0.25$ and $\alpha = 0.001$. Last value makes the algorithm work as an IBC procedure which eliminates exactly one center at a time. At the same time, the approach to the inclusion of elements of the ALA procedure in this greedy heuristic remains unchanged and the new algorithm with $\alpha = 0.001$ works slower than one with $\alpha = 0.25$ and still much faster than the IBC algorithm which uses Algorithm 4 for fitness function evaluation on each iteration. In addition, Table 1 shows results of thee simplified IBC method which evaluates the fitness function value without running ALA or another local search procedures. In fact. This simplified IBC procedure solves a $k$-medoid problem and then adjusts the result with ALA procedure. Such algorithm is entitled "IBC w/o local search". Such algorithm can be constructed after excluding Steps 3 and 4.5 from Algorithm 5 with $\alpha = 0.001$. If we remain $\alpha = 0.25$, i.e..we allow simultaneous eliminating several centers, we have a new deterministic algorithm entitled «IBC w/o local search, $\alpha = 0.25$». This version of the algorithm shows the best time and the worst preciseness.

For all inspected problems except those with the Jaccard metric, new algorithm shows the best results among all considered deterministic algorithms except IBC with local search which exceeds new algorithm by preciseness in several problems. However, new algorithm takes much less time. Results of new algorithm are less precise than the results of evolutionary algorithms. However, except problems with the Jaccard metric, the difference does not exceed

2.3% for problems with real data vectors and 3.8% for problems with Boolean data vectors.

Note that new algorithm has one important feature for solving a problem of automatic classification of the EEE components. Such problem is solved [3] as series of the k-means problems with $k = \overline{k_{min}, k_{max}}$ where $k_{min}$=1 (a single production batch without clusters assumed) and $k_{max}$ is chosen by a decider equal to some reasonable number. Algorithm 6 can be used for the k-means problem with $k$=$k_{max}$. Then, starting from Step 2, Algorithm 5 can run again for solving the succeeding problems until $k$=$k_{min}$. Thus, results for all values $k = \overline{k_{min}, k_{max}}$ can be calculated in a single run.

**Table 1.** Comparative results of new algorithm.

| Data set, quantity of data vectors, dimension, data type | Clusters q-ty $k$, metric, problem type | Algorithm | Result | Time, sec. | Std. devition of 10 runs |
|---|---|---|---|---|---|
| Chess (UCI), $N$=3197, $d$=50, Boolean | 50, $l_1$, k-сред-них | Local search multistart | 10020.2 | 35 | 19.11 |
| | | GA with fixed subsets recomb. | 9328.4 | 35 | 6.066 |
| | | GA with float. point alphabet | 9290 | 35 | 9.148 |
| | | IBC | - | - | - |
| | | IBC w/o local search | 9796 | 29.24 | |
| | | IBC w/o local search, $\alpha = 0.25$ | 10057 | 0.696 | Determ. |
| | | New algorithm, $\alpha = 0.25$ | 9649 | 0.694 | Determ. |
| | | New algorithm, $\alpha = 0.001$ | 9610 | 34.52 | Determ. |
| Examination of diodes, $N$=701, $d$=18,real | 30, $l_1$ нор-миро-ван., k-сред-них | Local search multistart | 5451.79 | 35 | 4,158 |
| | | GA with fixed subsets recomb. | 5427.32 | 35 | 6,31 |
| | | GA with float. Point alphabet | 5381.55 | 35 | 1,46 |
| | | IBC | 5390.12 | 2251.1 | Determ. |
| | | IBC w/o local search | 5494.05 | 0.686 | Determ. |
| | | IBC w/o local search, $\alpha = 0.25$ | 5483.89 | 0.111 | Determ. |
| | | New algorithm, $\alpha = 0.25$ | 5420.67 | 0.0669 | Determ. |
| | | New algorithm, $\alpha = 0.001$ | 5414.06 | 0.8962 | Determ. |
| Examination of diodes, $N$=701, $d$=18, real. | 30, $l_2^2$ nor-mali-zed., k-medo-ids | Local search multistart | 1950.52 | 7 | 5,362 |
| | | GA with fixed subsets recomb. | 1887.29 | 7 | 4,11 |
| | | GA with float. Point alphabet | 1885.79 | 7 | 1,457 |
| | | IBC | 1902.93 | 1062.1 | Determ. |
| | | IBC w/o local search | 1917.14 | 0.637 | Determ. |
| | | IBC w/o local search, $\alpha = 0.25$ | 1959.97 | 0.0324 | Determ. |
| | | New algorithm, $\alpha = 0.25$ | 1918.56 | 0.039 | Determ. |
| | | New algorithm, $\alpha = 0.001$ | 1912.43 | 0.6696 | Determ. |
| Examination of chip 1526LE2, $N$=620, $d$=120, real | 30, $l_2^2$ nor-mali-zed., k-means | Local search multistart | 12873.22 | 30 | 50,71 |
| | | GA with fixed subsets recomb. | 12536.19 | 30 | 3,11 |
| | | GA with float. Point alphabet | 12539.5 | 30 | 2,59 |
| | | IBC | 12556.04 | 4453.9 | Determ. |
| | | IBC w/o local search | 12727.13 | 0.5484 | Determ. |
| | | IBC w/o local search, $\alpha = 0.25$ | 12788.47 | 0.0088 | Determ. |

| | | | | | |
|---|---|---|---|---|---|
| | | New algorithm, $\alpha = 0.25$ | 12834.9 | 0.1258 | Determ. |
| | | New algorithm, $\alpha = 0.001$ | 12682.32 | 0.7528 | Determ. |

| Data set, quantity of data vectors, dimension, data type | Clusters q-ty *k*, metric, problem type | Algorithm | Result | Time, sec. | Std. devition of 10 runs |
|---|---|---|---|---|---|
| MissAmerica1 (UCI), *N*=6480, *d*=16, integer | 100, $l_2^2$, k-сред. | Local search multistart | 717488.7 | 60 | 1107.24 |
| | | GA with fixed subsets recomb. | 698055.6 | 60 | 460.62 |
| | | GA with float. point alphabet | 698054.5 | 60 | 406.42 |
| | | IBC | - | - | - |
| | | IBC w/o local search | 715762.6 | 159.8 | Determ. |
| | | IBC w/o local search, $\alpha = 0.25$ | 716150.6 | 1.85 | Determ. |
| | | New algorithm, $\alpha = 0.25$ | 707529.5 | 2.02 | Determ. |
| | | New algorithm, $\alpha = 0.001$ | 703786.4 | 199.5 | Determ. |
| PimaIndians Diabetes (UCI), *N*=768, *d*=9, real | 30, $l_2^2$, k-means | Local search multistart | 2072.425 | 36 | 2.4422 |
| | | GA with fixed subsets recomb. | 2045.196 | 36 | 1.7294 |
| | | GA with float. point alphabet | 2045.365 | 36 | 1.238 |
| | | IBC | 2043.481 | 1150.1 | Determ. |
| | | IBC w/o local search | 2096.06 | 0.81 | Determ. |
| | | IBC w/o local search, $\alpha = 0.25$ | 2144.771 | 0.031 | Determ. |
| | | New algorithm, $\alpha = 0.25$ | 2087.46 | 0.042 | Determ. |
| | | New algorithm, $\alpha = 0.001$ | 2056.35 | 1.058 | Determ. |
| Breast Cancer (UCI), *N*=699, *d*=10, cathegories | 20, Jaccard, k-medoids | Local search multistart | 184.1 | 5 | 0.8725 |
| | | GA with fixed subsets recomb. | 172.81 | 5 | 0.365 |
| | | GA with float. point alphabet | 172.62 | 5 | 0.0787 |
| | | IBC | 177.5 | 370.16 | Determ. |
| | | IBC w/o local search | 174.4 | 0.8898 | Determ. |
| | | IBC w/o local search, $\alpha = 0.25$ | 174.6 | 0.8072 | Determ. |
| | | New algorithm, $\alpha = 0.25$ | 175.2 | 0.437 | Determ. |
| | | New algorithm, $\alpha = 0.001$ | 175.7 | 0.7352 | Determ. |
| Zoo (UCI), *N*=101, *d*=10, cathegories | 10, метрика Жаккара, k-medoids | Local search multistart | 6.4285 | 1 | 0.0287 |
| | | GA with fixed subsets recomb. | 6.4118 | 1 | 0 |
| | | GA with floating point | 6.4118 | 1 | 0 |
| | | alphabet | 6.4706 | 4.533 | Determ. |
| | | IBC | 6.4706 | 0.015 | Determ. |
| | | IBC w/o local search | 6.4706 | 0.010 | Determ. |
| | | IBC w/o local search, $\alpha = 0.25$ | 6.4706 | 0.009 | Determ. |
| | | New algorithm, $\alpha = 0.25$ | 6.4706 | 0.012 | Determ. |
| | | New algorithm, $\alpha = 0.001$ | | | |
| Ionosphere (UCI), *N*=351, *d*=35, real. | 10, $L_1$, k-means | Local search multistart | 2530.40 | 4 | 2.117 |
| | | GA with fixed subsets recomb. | 2526.77 | 4 | 0.0257 |
| | | GA with float. point alphabet | 2527.00 | 4 | 0.0082 |
| | | IBC | 2536.06 | 76.13 | Determ. |

| | | | | | |
|---|---|---|---|---|---|
| | | IBC w/o local search | 2546.54 | 0.1124 | Determ. |
| | | IBC w/o local search, $\alpha = 0.25$ | 2572.40 | 0.0198 | Determ. |
| | | New algorithm, $\alpha = 0.25$ | 2531.03 | 0.025 | Determ. |
| | | New algorithm, $\alpha = 0.001$ | 2533.31 | 0.151 | Determ. |

## Conclusion

Proposed algorithm allows solving k-means and k-medoids problems in appropriate time. Achieved results are insignificantly less precise than the results of the evolutionary algorithms. However, new algorithm is deterministic and this fact makes its results easy for checking and interpreting by all concerned parts. The preciseness and effectiveness of the program realization of new algorithm allow solving the problem of classifying the EEE components by production batches based on the quality examination test data.

## References

[1] Fedosov V V and Orlov V I 2011 Minimal necessary extent of examination of microelectronic products at inspection test stage *Izvestiya Vuzov. Priborostroenie* **54(4)** 68–62

[2] Kharchenko V S and Yurchenko Yu B 2003 Rating of fault-tolerant onboard complexes frames at usage electronic components industry *Tekhnologiya I konstruirovanie v elektronnoy apparature* **2** 3–10

[3] Kazakovtsev L A, Orlov V I, Stupina A A and Masich I S 2014 Problem of electronic components classifying *Vestnik SibGAU* **4(56)** pp 55-61

[4] Ackermann M R et al 2012 StreamKM: A Clustering Algorithm for Data Streams *J. Exp. Algorithmics* **17** 2.4:2.1-2.30

[5] Kanungo T, Mount D M, Netanyahu N S, Piatko C D, Silverman R and Wu A Y 1999 Computing nearest neighbors for moving points and applications to clustering *Proc. of the tenth annual ACM-SIAM symp. on Discrete algorithms* (Society for Industrial and Applied Mathematics) pp 931-932

[6] Kazakovtsev L A, Stupina A A and Orlov V I 2014 Modification of the genetic algorithm with greedy heuristic for continuous location and classification problems *Sistemy upravleniya i informatsionnye tekhnologii* **2(56)** 31–34

[7] Hamiter L 1991 The History of Space Quality EEE Parts in the United States *ESA Electronic Components Conf., ESTEC (Noordwijk, The Netherlands, 12–16 Nov 1990 ESA SP-313)*

[8] Kirkconnell C S et al 2014 High Efficiency Digital Cooler Electronics for Aerospace Applications *Proc. SPIE 9070, Infrared Technology and Applications XL 90702Q (June 24, 2014)*

[9] Farahani R Z and Hekmatfar M editors 2009 *Facility Location: Concepts, Models, Algorithms and Case Studies* (Berlin Heidelberg: Springer-Verlag)

[10] Kazakovtsev L A and Stupina A A 2014 Fast Genetic Algorithm with Greedy Heuristic for p-Median and k-Means *Problems IEEE 2014 6th Int. Congress on Ultra Modern Telecommunications and Control Systems and Workshops ICUMT (St.-Petersburg)* pp 702-706

[11] Kazakovtsev L A and Antamoshkin A N 2014 Genetic Algorithm wish Fast Greedy Heuristic for Clustering and Location Problems *Informatica* **38(3)** 229-240

[12] Deza M M and Deza E 2009 *Encyclopedya of Distances* (Berlin Heidelberg: Springer-Verlag)

[13] Trubin V A 1978 An efficient algorithm for the Weber problem with a rectangular metric *Kibernetika* **6** 67-70
[14] Kaufman L and Rousseeuw P J 1990 *Finding Groups in Data: an Introduction to Cluster Analysis* (New York: John Wiley & Sons)
[15] Park H S and Jun C-H 2009 A simple and fast algorithm for K-medoids clustering *Expert Systems with Applications* **36** 3336–41
[16] Kuehn A A and Hamburger M J 1963 A heuristic program for locating warehouses *Management Science* **9(4)** 643-666
[17] Lucasius C B, Dane A D and Kateman G 1993 On K-Medoid Clustering of Large Data Sets with the Aid of a Genetic Algorithm: Background, Feasibility and Comparison *Analytical Chimica Acta* **282** 647-669
[18] Zhang Q and Couloigner I 2005 A New Efficient K-Medoid Algorithm for Spatial Clustering *ICCSA 2005, LNCS 3482* pp 181–189
[19] Sheng W and Liu X 2004 A Genetic K-Medoids Clustering Algorithm *Journal of Heuristics* **12(6)** 447-466
[20] Cooper L 1963 Location-allocation problem *Oper. Res.* **11** 331–343
[21] Arthur D and Vassilvitskii S 2007 k-Means++: the Advantages of Careful Seeding *Proc. of the eighteenth annual ACM-SIAM symp. on Discrete algorithms* pp 1027–1035
[22] Mishra N, Oblinger D and Pitt L 2001 Sublinear time approximate clustering *12th SODA* pp 439–447
[23] Sun Zh, Fox G, Gu W and Li Zh 2014 A parallel clustering method combined information bottleneck theory and centroid-based clustering *The Journal of Supercomputing* **69(1)** 452–467
[24] Alp O, Erkut E and Drezner Z 2003 An Efficient Genetic Algorithm for the p-Median Problem *Annals of Operations Research* **122** 21–42
[25] Neema M N, Maniruzzaman K M and Ohgai A 2011 New Genetic Algorithms Based Approaches to Continuous p-Median Problem *Netw. Spat. Econ.* **11** 83–99
[26] Patrick M, Murphy P M and Aha D W 1994 UCI Repository of Machine Learning Databases http://www.cs.uci.edu/mlearn/mlrepository.html

IOP Conf. Series: Materials Science and Engineering **94** (2015) 012016 doi:10.1088/1757-899X/94/1/012016

# DETERMINISTIC ALGORITHM WITH AGGLOMERATIVE HEURISTIC FOR LOCATION PROBLEMS*

**L Kazakovtsev[1] and A Stupina[2]**

[1] PhD in Engineering, Assoc. Professor Siberian State Aerospace University named after Academician M.F. Reshetnev
31 “Krasnoyarskiy Rabochiy” prospect, Krasnoyarsk, 660037, Russia.
[2] Professor Siberian State Aerospace University named after Academician M.F. Reshetnev
31 “Krasnoyarskiy Rabochiy” prospect, Krasnoyarsk, 660037, Russia.
Professor Siberian State University, 3 Vuzovskii, Krasnoyarsk, 660025, Russia

E-mail: h677hm@gmail.com

**Abstract.** Authors consider the clustering problem solved with the k-means method and p-median problem with various distance metrics. The p-median problem and the k-means problem as its special case are most popular models of the location theory. They are implemented for solving problems of clustering and many practically important logistic problems such as optimal factory or warehouse location, oil or gas wells, optimal drilling for oil offshore, steam generators in heavy oil fields. Authors propose new deterministic heuristic algorithm based on ideas of the Information Bottleneck Clustering and genetic algorithms with greedy heuristic. In this paper, results of running new algorithm on various data sets are given in comparison with known deterministic and stochastic methods. New algorithm is shown to be significantly faster than the Information Bottleneck Clustering method having analogous preciseness.

The *k*-means problem can be classified as a continuous problem of the location theory [1, 2, 3]. The aim is to find *k* points (centers, centroids, medoids) in a *d*-dimensional space such that the total distance from each of the data vectors (known points, measurement result vectors) to the nearest of *k* chosen centers reaches its minimum:

$$F(X_1,\dots,X_k)=\sum_{i-1}^{N} min_{x\in\{x_1,\dots,x_k\}} w_i\|A_i - X\|_2^2. \qquad (1)$$

Here, $w_i$ are nonnegative weight coefficients. If all weight coefficient are equal to 1 then we have a k-means problem.

Euclidean metric is commonly used for logistic problems concerned with real objects location on a plane or in a 3D space. Such objects are factories, warehouses [1], gas or oil wells [4, 5, 6], stean generators in heavy oil fields [7], transportation or telecommunication nodes [1].

In case of clustering problems, the squared Euclidean norm is most popular distance metric

* Results were obtained in the framework of the state task № 346 of the Ministry of Education and Science of the Russian Federation

used for calculating differences (distances) in a normalized space of characteristics [1]. Using the rectilinear (Manhattan) [8] norm as a distance metric in the *k*-means model allows to reach results of the same precision as the precision of data vectors. In this case, the value of each coordinate of the result coincides with the value of the same coordinate of one of the data vectors [1, 9]. Moreover, the result of the *k*-means problem rectilinear metric are more stable under the influence of the outliers in the data which exist due to measurement errors and defective components in the lot. Other approach which allows to achieve the results of the same precision as data vectors is solving the *k*-medoid problem [8, 9]. In this problem, the cluster centers which minimize the total distance are searched among the data vectors only.

The *k*-means method uses the ALA procedure (Alternating Location-Allocation) which includes two simple steps:

**Algorithm 1.** ALA procedure.

Required: data vectors $A_1 \dots A_N$, $k$ initial cluster centers $X_1 \dots X_k$.

1. For each center $X_i$, determine its cluster $C_i$ as a subset of the data vectors for which this center $X_i$ is the closest one.

2. For each cluster $C_i$, recalculate its center $X_i$.

$$X_i = \arg\min_{X} \sum_{y \in C_i} w_i \|x\text{-}y\|.$$

3. Repeat Step 1 unless Steps 1, 2 made no change in data.

Traditional usage of the *k*-means method with squared Euclidean metric ($l_2$) has one important advantage: in this case, calculating a center of a cluster is a simple problem solved in one iteration as calculating the mean value for each coordinate of all data vectors in the cluster. These mean values are coordinates of the new cluster center [1]. If the *i*-th cluster center $X_i=(x_{i,1},\dots,x_{i,d})$ is a vector in a *d*-dimensional space and data vectors $A_j=(a_{i,1},\dots,a_{j,d})$, $j=\overline{1,N}$ also have *d* dimensions then the new cluster center for a k-means problem with squared Euclidean metric is calculated as [1]:

$$x'_{i,k} = \sum_{y \in C_i} y_k / |C_i|, k = \overline{1,d}\,.$$

In the ALA procedure with the rectilinear metric, each coordinate of the cluster center is calculated as the median value of this coordinate among all data vectors which belong to the cluster. This process can be described as follows.

**Algorithm 2.** Calculating *i*-th cluster center (median) in case of the $l_1$ metric.

1. For each $k=\overline{1,d}$ do:

1.1. Sort vectors $A_i = (a_{j,1},\dots a_{j,d}) \in C_i$ in acceding order of the *k*-th coordinate, form a sequence $a'_{1,k},\dots,a'_{|C_i|,k}$. Here, $|C_i|$ is the power of the set (cluster).

1.2. Calculate $m = \left[ |C_i| / 2 \right]$. Store $x'_{i,k} = a'_{m,k}$. Here, $[\bullet]$ is the integer part.

1.3. Next iteration 1.

2. Return $X'_i = (x'_{i,1},\dots,x'_{i,d})$

This algorithm returns a value of new center which has values of each coordinate coinciding with a coordinate of some data vector.

In the case of the *k*-medoid problem, procedure of determining of each cluster center requires the exhaustive search among all data vectors in the cluster. However, many researchers propose faster analogous local search procedures [11, 12, 13] which do not guarantee an exact solution.

Except special cases, the *k*-means and *k*-medoids problems are *NP* hard and require global search [14].

The result of the ALA procedure depends on the choice of the initial cluster centers. Known *k*-means++ algorithm [15] has an advantage in comparison with the chaotic choice of the initial centers and guarantees the statistical preciseness of the result $O(\log(p))$. However, such preciseness is insufficient for many practically important problems. For such cases, researchers propose various recombination techniques for initial center sets [1].

The ALA procedure can be optimized with use of many techniques. For example, sampling procedures [16] solve the *k*-means problem for the randomly selected subset of the data vectors and use the achieved result as an initial set of centers for solving the original problem. Authors propose various algorithms for streaming data processing [17] applicable for big data analysis and many other methods.

The dependence of the results of the ALA procedure on the initial centers seeding is a serious problem for the reproducibility of the classification algorithm results: depending on the initial centers seeding, different algorithm starts classify the same data vectors as elements of various clusters. Thus, an algorithm for solving *k*-means problem which returns a stable result is preferred.

The Information Bottleneck Clustering method (IBC) is a deterministic method for solving the cluster analysis and classification problems able to achieve perfect results in many cases [18]. This algorithm starts from considering each data vector as a separate cluster. Then, clusters are removed one-by-one until the desired clusters quantity remains. Each time, the algorithm eliminates such cluster that its elimination gives the smallest increment of the objective function value. For the *k*-means problem, this algorithm eliminates the cluster center which gives the smallest total distance from data vectors to the closest remaining centers. Such algorithms are extremely slow [18]. The genetic algorithms with greedy heuristic initially designed for the discrete *k*-median problem on a network [19] are compromise variants. However, they are not determined algorithms. In [20], author proposes an approach for adaptation of these algorithms for the continuous location problems. The idea of such approach can be described as follows [17].

**Algorithm 3.** Genetic algorithm with greedy heuristic for *k*-median problems.

Required: data vectors $A_1 \ldots A_N$, population size $N_p$.

1. Form (randomly or using the *k*-means++ procedure) $N_p$ various initial solutions $\chi_1, \ldots \chi_{N_p} \subset \{\overline{1,N}\}$, $|\chi_i| = k \forall i = \overline{1,N_p}$. Each of such solutions is a set of $k$ data vectors used as the initial solutions of the ALA procedure. Calculate the fitness function value (total distance) $F_{fitness}(\chi)$ for each of the initial solutions using Algorithm 3 and store the fitness function values to $f_1, \ldots, f_{N_p}$.

2. If the stop conditions are satisfied then STOP. Return solution $\chi_{i^*}$, with the smallest value $f_i$. For the final solution, the ALA procedure (Algorithm 1) runs.

3. Choose randomly two indexes $k_1, k_2 \in \{\overline{1,N}\}, k_1 \neq k_2$.

4. Form an interim solution $\chi_c = \chi_{k_1} \cup \chi_{k_2}$.

5. If $|\chi_c| > k$ then go to Step 7:

6. Calculate $j^* = \arg\min_{j \in \chi_c} F_{fitness}(\chi_c \setminus \{j\})$. Exclude $j^*$ from $\chi_c : \chi_c = \chi_c \setminus \{j^*\}$. Go to Step 5.

7. If $\exists i\in\{\overline{1,N_p}\}:\chi_i=\chi_c$ then go to Step 2.

8. Choose index $k_3\in\{\overline{1,N_p}\}$. First, two indexes $k_4,k_5\in\{\overline{1,N}\}$ are chosen randomly, then if $f_{k_4}>f_{k_5}$ then $k_3=k_4$ else $k_3=k_5$.

9. Replace $\chi_{k_3}$ and the corresponding fitness function value: $\chi_{k_3}=\chi_c$, $f_{k_3}=F_{fitness}(\chi_c)$. Go to Step 2.

Steps 5-6 of this algorithm realize the greedy heuristic, the successive elimination of the centers from an interim solution.

An analogous heuristic was proposed by Kuehn and Hamburger in 1963 [21]. The IBC method is based on the same principle of the successive elimination of the clusters from an interim solution [22]. Both method of Kuehn and Hamburger and the IBC chose an unfeasible solution coinciding with the whole data vectors set as the initial solution.

The fitness function for the $k$-means problem can be calculated for an initial or interim solution as follows:

**Algorithm 4.** Calculating fitness function value $F_{fitness}(\chi)$.

Required: solution $\chi$.

1. Run Algorithm 1 with initial centers set $\{A_i|i\in\chi\}$ resulting with centers set $\{X_1,...,X_p\}$.
2. Return $F_{fitness}(\chi)=\sum_{i=1}^{N} w_i \min_{j\in\{1,k\}} L(X_j,A_i)$.

This algorithm is a computationally intensive procedure. Other approach [20] is based on the immediate usage of the total distance from data vectors to the closest center in the interim solution as the fitness function value for Algorithm 3. In this case, Step 1 of Algorithm 4 is omitted. In fact, this approach solves a $k$-medoid problem and adjusts the solution with the ALA procedure at the final iteration of the greedy heuristic. Such approach is much faster. However, it reduces the preciseness.

An advantage of the IBC method is its determinancy. This method does not use any random values and each start of the algorithm results in the same set of cluster centers. The exact reproduction of the results is impossible when using algorithms which include random search elements. The IBC method slows down the calculation.

In [3], authors propose a modification of the greedy heuristic used in Step 6 of Algorithm 3. This method uses points in the $d$-dimensional space instead of data vector indexes as an alphabet of the genetic algorithm. Such points represent the cluster centers. Authors entitle this modification Algorithm with Floating Point Alphabet. Results of the genetic algorithm with this heuristic are much more precise than results of the modification described in [20] (in [20], authors propose solving a $k$-medoid problem). At the same time, the iterations in Algorithm 3 with this heuristic are performed much faster than Algorithm 4 as the greedy heuristic. Moreover, decrease of the computational complexity of Step 6 of Algorithm 3 makes solving large-scale problems possible. In [3], authors report results of solving problems with up to 560000 data vectors. Such heuristic can be described as follows.

**Algorithm 5.** Greedy heuristic for the GA with floating point alphabet (used instead of Steps 4-7 of Algorithm 3).

Required: set of data vectors $V=(A_1,...,A_N)\in\mathrm{R}^d$, number $k$ of clusters, two "parent" center sets $\chi_{k_1}$ and $\chi_{k_2}$, parameter $\alpha$.

1. Form interim solution $\chi_c=\chi_{k_1}\cup\chi_{k_2}$. Run ALA algorithm from initial solution $\chi_c$. Store its result to $\chi_c$.

2. If $|\chi_c|=k$ then start ALA procedure from initial solution $\chi_c$, STOP and return $\chi_c$.

2.1. Calculate distances to the closest element of $\chi_c$:

$$d_i=\min_{X\in\chi_c} L(X,A_i)\,\forall i=\overline{1,N}.$$

Form clusters around each center in $\chi_c$. $C_i=\arg\min_{X\in\chi_c} L(X,A_i)\,\forall i=\overline{1,N}.$

Calculate distances from each data vector to the second closest center in $\chi_c$.

$$d_i=\min_{Y\in\chi_c\setminus\{C_i\}} L(Y,A_i)\,\forall i=\overline{1,N}.$$

3. For each center $X\in\chi_c$, calculate $\delta_X=F(\chi_c\setminus\{X\})=\sum_{i:X\in C_i}(D_i-d_i)$.

4.1. Calculate $n_\delta=max\{[\alpha(|\chi_c|-k)],1\}$. Sort values $\delta_X$ in acceding order and choose a subset $\chi_{e\,lim}=\{X_1,...,X_{n_\delta}\}$ of $n_\delta$ data vectors with minimal values $\delta_X$.

4.2. For each $j\in\{\overline{2,|\chi_{e\,lim}|}\}$do: if $\exists q\in\{\overline{1,(j-1)}\}$: $L(X_j,X_q)<L_{min}$ then eliminate $X_j$ from set $\chi_{e\,lim}$. Here, $L_{min}=\min_{X\in\chi_c}\{max\{L(X,X_j),L(X,X_q)\}\}$.

4.3. Set $\chi_c=\chi_c\setminus\chi_{e\,lim}$.

4.4. Reallocate data vectors to the closest centers: $C_i^*=\arg\min_{X\in\chi_c} L(X,A_i)\,\forall i=\overline{1,N}.$

4.5. For each $X\in\chi_c$ if $\exists i\in\{\overline{1,N}\}$: $C_i=X$ and $C_i^*\neq X$ then recalculate center $X^*$ of cluster $C_X^{clust}=\{A_i\,|\,C_i^*=X,i=\overline{1,N}\}$. Store $\chi_c=(\chi_c\setminus\{X^*\})\cup\{X\}$.

5. Go to Step 2.

Value $\alpha$ is an important parameter of this heuristic. This value determines the percentage of the superfluous cluster centers eliminated in a single iteration. Authors propose value 0.2. Bigger values make the algorithm run faster and reduce its preciseness. Small values $\alpha$ make it work as Algorithm 4 and eliminate a single center at each iteration. We used $\alpha$ =0.25.

This heuristic combines the greedy heuristic [19, 20, 23] with elements of the modified ALA procedure performed at each iteration and allows to eliminate up to 20-25% superfluous cluster centers until the required quantity of centers remains. Algorithm 4 requires performing $p(k_0\text{-}k)$ runs of the ALA procedure (here, $k_0$ is the initial centers quantity). Algorithm 5 reduces quantity of iterations down to $O(\log(k_0\text{-}k))$. Moreover, each iteration does not require performint the whole ALA procedure. Instead, its separate optimized elements (location – allocation) are performed.

It can be easily seen that if $k_0=N$ then both Algorithm 5 and IBC method start their initial iterations analogously: number of cluster centers coincides with the number of data vectors. Moreover, if $k_0=N$ then choosing th initial centers is not random: all data vectors are chosen as the initial centers. Thus, the following deterministic algorithm can be proposed.

**Algorithm 6.** New deterministic greedy heuristic algorithm.

Required: set of data vectors $V=(A_1,...,A_N)\in\mathrm{R}^d$, number $k$ of clusters, parameter $\alpha$.

1. Set $\chi_c=V$.

2. Run Algorithm 5 starting from Step 2.

Such algorithm was successfully applied for middle-scale problems, up to *N*=6500 data vectors. The results are gathered in Table 1. We used example problems from the UCI library [23] and problems with real data of EEE components examination [1]. For small-scale problems, the results are shown in comparison with the IBC method, genetic algorithm with greedy heuristic and genetic algorithm with recombination of fixed length subsets [24]. We used various distance metrics. In addition, *k*-medoid problems [9, 10] were solved. For several problems, the results of new algorithm are insignificantly worse than the results of the IBC. At the same time, time needed for problem solution reduces many times. In Table 1, all results of new algorithm are shown for $\alpha = 0.25$ and $\alpha = 0.001$. Last value makes the algorithm work as an IBC procedure which eliminates exactly one center at a time. At the same time, the approach to the inclusion of elements of the ALA procedure in this greedy heuristic remains unchanged and the new algorithm with $\alpha = 0.001$ works slower than one with $\alpha = 0.25$ and still much faster than the IBC algorithm which uses Algorithm 4 for fitness function evaluation on each iteration. In addition, Table 1 shows results of the simplified IBC method which evaluates the fitness function value without running ALA or another local search procedures. In fact, this simplified IBC procedure solves a *k*-medoid problem and then adjusts the result with ALA procedure. Such algorithm is entitled "IBC w/o local search". Such algorithm can be constructed after excluding Steps 3 and 4.5 from Algorithm 5 with $\alpha = 0.001$. If we remain $\alpha = 0.25$, i.e. we allow simultaneous eliminating several centers, we have a new deterministic algorithm entitled "IBC w/o local search, $\alpha = 0.25$". This version of the algorithm shows the best time and the worst preciseness.

For all inspected problems except those with the Jaccard metric, new algorithm shows the best results among all considered deterministic algorithms except IBC with local search which exceeds new algorithm by preciseness in several problems. However, new algorithm takes much less time. Results of new algorithm are less precise than the results of evolutionary algorithms. However, except problems with the Jaccard metric, the difference does not exceed 2.3% for problems with real data vectors and 3.8% for problems with Boolean data vectors.

Note that new algorithm has one important feature for solving a problem of automatic classification of the EEE components. Such problem is solved [23] as series of the k-means problems with $k = \overline{k_{min}, k_{max}}$ where $k_{min}$=1 (a single production batch without clusters assumed) and $k_{max}$ is chosen by a decider equal to some reasonable number. Algorithm 6 can be used for the k-means problem with $k=k_{max}$. Then, starting from Step 2, Algorithm 5 can run again for solving the succeeding problems until $k=k_{min}$. Thus, results for all values $k = \overline{k_{min}, k_{max}}$ can be calculated in a single run.

**Table 1.** Comparative results of new algorithm

| Data set, quantity of data vectors, dimension, data type | Clusters q-ty *k*, metric, problem type | Algorithm | Result | Time, sec. | Std. devition of 10 runs |
|---|---|---|---|---|---|
| Chess (UCI), | 50, $l_1$, | Local search multistart | 10015.89 | 30 | 19.08 |
| *N*=3197, *d*=50, | k- means | GA with fixed subsets recomb. | 9331.8 | 30 | 6.074 |
| Boolean | | GA with float. point alphabet | 9286 | 30 | 9.135 |
| | | IBC | - | - | - |
| | | IBC w/o local search | 9796 | 29.18 | |
| | | IBC w/o local search, $\alpha = 0.25$ | 10057 | 0.688 | Determ. |
| | | New algorithm, $\alpha = 0.25$ | 9649 | 0.696 | Determ. |

| | | | | | |
|---|---|---|---|---|---|
| | | New algorithm, $\alpha = 0.001$ | 9610 | 34.58 | Determ. |
| MissAmerica1 (UCI), $N$=6480, $d$=16, integer | 100, $l_2^2$, k- means | Local search multistart | 717869.5 | 60 | 1105.17 |
| | | GA with fixed subsets recomb. | 698188.2 | 60 | 455.64 |
| | | GA with float. point alphabet | 698134.6 | 60 | 408.36 |
| | | IBC | - | - | - |
| | | IBC w/o local search | 715762.6 | 157.6 | Determ. |
| | | IBC w/o local search, $\alpha = 0.25$ | 716150.6 | 1.77 | Determ. |
| | | New algorithm, $\alpha = 0.25$ | 707529.5 | 2.18 | Determ. |
| | | New algorithm, $\alpha = 0.001$ | 703786.4 | 197.8 | Determ. |
| PimaIndians Diabetes (UCI), $N$=768, $d$=9, real | 30, $l_2^2$, k-means | Local search multistart | 2071.424 | 40 | 2.3312 |
| | | GA with fixed subsets recomb. | 2046.158 | 40 | 1.7294 |
| | | GA with float. point alphabet | 2044.143 | 40 | 1.238 |
| | | IBC | 2043.481 | 1149 | Determ. |
| | | IBC w/o local search | 2096.06 | 0.75 | Determ. |
| | | IBC w/o local search, $\alpha = 0.25$ | 2144.771 | 0.029 | Determ. |
| | | New algorithm, $\alpha = 0.25$ | 2087.46 | 0.038 | Determ. |
| | | New algorithm, $\alpha = 0.001$ | 2056.35 | 1.061 | Determ. |
| Breast Cancer (UCI), $N$=699, $d$=10, cathegories | 20, Jaccard, k-medoids | Local search multistart | 184.2 | 7 | 0.8712 |
| | | GA with fixed subsets recomb. | 172.79 | 7 | 0.298 |
| | | GA with float. point alphabet | 172.61 | 7 | 0.0815 |
| | | IBC | 177.5 | 369.08 | Determ. |
| | | IBC w/o local search | 174.4 | 0.8886 | Determ. |
| | | IBC w/o local search, $\alpha = 0.25$ | 174.6 | 0.8068 | Determ. |
| | | New algorithm, $\alpha = 0.25$ | 175.2 | 0.441 | Determ. |
| | | New algorithm, $\alpha = 0.001$ | 175.7 | 0.736 | Determ. |
| Zoo (UCI), $N$=101, $d$=10, cathegories | 10, Jac card, k-medoids | Local search multistart | 6.4261 | 1 | 0.0278 |
| | | GA with fixed subsets recomb. | 6.4116 | 1 | 0 |
| | | GA with floating point alphabet | 6.4116 | 1 | 0 |
| | | IBC | 6.4706 | 4.529 | Determ. |
| | | IBC w/o local search | 6.4706 | 0.013 | Determ. |
| | | IBC w/o local search, $\alpha = 0.25$ | 6.4706 | 0.015 | Determ. |
| | | New algorithm, $\alpha = 0.25$ | 6.4706 | 0.011 | Determ. |
| | | New algorithm, $\alpha = 0.001$ | 6.4706 | 0.009 | Determ. |
| Ionosphere (UCI), $N$=351, $d$=35, real. | 10, $L_1$, k-means | Local search multistart | 2531.68 | 6 | 2.117 |
| | | GA with fixed subsets recomb. | 2526.55 | 6 | 0.0265 |
| | | GA with float. point alphabet | 2527.01 | 6 | 0.0081 |
| | | IBC | 2536.06 | 77.01 | Determ. |
| | | IBC w/o local search | 2546.54 | 0.1132 | Determ. |
| | | IBC w/o local search, $\alpha = 0.25$ | 2572.40 | 0.021 | Determ. |
| | | New algorithm, $\alpha = 0.25$ | 2531.03 | 0.0199 | Determ. |
| | | New algorithm, $\alpha = 0.001$ | 2533.31 | 0.148 | Determ. |

**Conclusion.** Proposed algorithm allows solving *k*-means and *k*-medoids problems in appropriate time. Achieved results are insignificantly less precise than the results of the evolutionary algorithms. However, new algorithm is deterministic and this fact makes its results easy for checking and interpreting by all concerned parts.

**References**

[1] Farahani R Z and Hekmatfar M editors 2009 Facility Location Concepts, Models, Algorithms and Case Studies, *Springer-Verlag Berlin Heidelberg.*

[2] Kazakovtsev L A, Stupina A A 2014 Fast Genetic Algorithm with Greedy Heuristic for p-Median and k-Means Problems *IEEE 2014 6th International Congress on*

*Ultra Modern Telecommunications and Control Systems and Workshops (ICUMT). St.-Petersburg, October 6-8* pp.702-706.

[3] Kazakovtsev L A, Antamoshkin A N 2014 Genetic Algorithm wish Fast Greedy Heuristic for Clustering and Location *Problems Informatica* **38(3)** pp. 229-240.

[4] Eiselt H A 1992 Location Modelling in Practice *American Journal of Mathematical and Management Sciences* **12(1)** pp. 3-18, DOI: 10.1080/01966324.1992.10737322

[5] Kabadi S, Murty K G, Spera C 1996 Clustering Problems in Optimization Models *Computational Economics* **9** pp.229-239

[6] Church R L, Scaparra M P, Middleton R S 2004 Identifying Critical Infrastructure: The Median and Covering Facility Interdiction Problems *Annals of the Association of American Geographers* **94(3)** pp.491-502.

[7] Rosing K E 1992 The optimal location of steam generators in large heavy fields *American Journal of Mathematical and Management Sciences* **12** pp.19-42.

[8] Deza M M and Deza E 2009 Encyclopedya of Distances *Springer Verlag, Berlin, Heilderberg.*

[9] Trubin V A 1978 Effektivnyy algoritm dlya zadachi Vebera s pryamougol'noy metrikoy *Kibernetika* **6** pp.67-70, DOI:10.1007/BF01070282/

[10] Kaufman L and Rousseeuw P J 1990 *Finding Groups in Data: an Introduction to Cluster Analysis. New York: John Wiley & Sons.*

[11] Park H S, Jun C-H 2009 A simple and fast algorithm for K-medoids clustering *Expert Systems with Applications* **36** pp.3336–3341

[12] Lucasius C B, Dane A D and Kateman G 1993 On K-Medoid Clustering of Large Data Sets with the Aid of a Genetic Algorithm: Background, Feasibility and Comparison *Analytical Chimica Acta* **282** pp.647-669.

[13] Zhang Q, Couloigner I 2005 A New Efficient K-Medoid Algorithm for Spatial Clustering *ICCSA 2005, LNCS* **3482** pp. 181–189.

[14] Cooper L 1963 Location-allocation problem *Oper. Res.* **11** pp. 331–343

[15] Arthur D and Vassilvitskii S 2007 k-Means++: the Advantages of Careful Seeding *Proceedings of the eighteenth annual ACM-SIAM symposium on Discrete algorithms* pp. 1027–1035

[16] Mishra N, Oblinger D, Pitt L 2001 Sublinear time approximate clustering *12th SODA* pp. 439–447

[17] Ackermann M R et al. 2012 StreamKM: A Clustering Algorithm for Data Streams *J. Exp. Algorithmics* vol.17 article 2.4, DOI: 10.1145/2133803.2184450/

[18] Sun Zh, Fox G, Gu W, Li Zh 2014 A parallel clustering method combined information bottleneck theory and centroid-based clustering *The Journal of Supercomputing* **69(1)** pp. 452–467. DOI: 10.1007/s11227-014-1174-1.

[19] Alp O, Erkut E, Drezner Z 2003 An Efficient Genetic Algorithm for the p-Median *Problem Annals of Operations Research* **122(1–4)** pp. 21–42

[20] Neema M N, Maniruzzaman K M, Ohgai A 2011 New Genetic Algorithms Based Approaches to Continuous p-Median *Problem Netw. Spat. Econ.* **11** pp.83–99. DOI: 10.1007/s11067-008-9084-5

[21] Kuehn A A, Hamburger M J 1963. A heuristic program for locating warehouses *Management Science* **9(4)** pp.643-666

[22] Kazakovtsev L A, Orlov V I, Stupina A A, Masich I S 2014 Problem of electronic components classifying *Vestnik SibGAU* **4(56)** pp.55-61.

[23] Patrick M Murphy P M, Aha D W 1994 UCI Repository of Machine Learning Databases. URL http://www.cs.uci.edu/ mlearn/mlrepository.html (accessed 02.01.2015)
[24] Sheng W, Liu X 2004 A Genetic K-Medoids Clustering Algorithm *Journal of Heuristics* **12(6)** pp. 447-466, DOI: 10.1007/s10732-006-7284-z

# INDEXES SYSTEM OF TECHNOLOGICAL CONDITION ASSESSMENT OF ECONOMIC BRANCHES

**M N Chuvashova, N T Avramchikova, A N Antamoshkin**
Siberian State Aerospace University named after Academician M.F. Reshetnev
31 Krasnoyarsky Rabochy Av., Krasnoyarsk, 660037, Russian Federation

E-mail: nikolesole@yandex.ru

**Abstract.** The increased level of innovative production process, connected with the current trends, points out the necessity of economic diversification of the whole national economy as well as regional economies in order to increase competitiveness and stable development. Russian regional economies are characterized with local directive of development and innovative processes have evident local vector. Intensive development of Siberian regional economies, which depends on oil and mining industries, considerably falls behind the world indicators according to the GRP output per head. To improve the quality of economic space the authors have suggested a new scientific approach, which allows qualitative assessment inside the economic space of resource-based regions, based on principles of high technological modes development inside economic branches taking into account density, regular enterprise distribution and connectivity of commercial organizations as well as secures innovative development of regional economy and its competitiveness. In this context it is necessary to develop a modern system of indexes, characterizing the structure of economic branches in accordance with present technological modes and at the same time the dynamics of appropriate structural shifts in regional economies of this type.

Economic growth in Russia is mainly formed due to mining and processing industry which provides over half of all the budget revenues of the country.

Emerging of economy at the cost of primary sector determines extensive rather than intensive modification of the development of the country. High prices on world commodity markets in primary resources have made a tangible contribution to the maintenance of high rate in GDP growth.

The decline and subsequent shutdown of resource-intensive productions in the 90s had a positive impact on the decrease in unit costs of product release of the industrial sector as a whole, having enhanced the efficiency in the use of resources. This situation had a short-term effect and has led to the generation of appalling records in the costs of resources for the production release in the process of demand recovery. It was partly achieved due to the increase in the rate of the load of manufacturing facilities close to the optimum level (75 – 85%).The average age of equipment in the industrial sector has risen from 8.5 to 22 years which practically means the need to replace more than 40% of the total volume of basic production assets. GDP growth rates data published in the press reflect quantitative changes taking place in terms of extensive as well as intensive constituents of the economic growth [1].

By the end of 2013 the deterioration rate of basic assets had been equal to 48.6%in the whole country, 53.7% for enterprises of mining industry and 47.8% for manufacturing industries [2]. The existence of physically and mentally worn-out basic assets results in inefficiency of their use and insufficient rates of renewal restrain the modernization of the economy as a whole.

The current management techniques, which develop regional economic and social systems, are accomplished with the application of official data taken from statistical records containing a limited number of indicators, which define the technological condition of regional economy branches [3]. The relevance of present inquiries is emphasized in the President's Message to the Federal Assembly on 12th December 2013 devoted to the 20th anniversary of Russian Federation Constitution: "… I propose to establish the statistical estimation system of the technological condition level in industry sectors in order to have an objective picture of our competitiveness" [4].

The establishment of regional economic space assessment system requires selection of appropriate national measures, introduction of some new measures and their distribution based on economic space assessment criteria.

According to up-to-date theoretical principles of regional economy the regional economic space quality is identified with the following characteristics: density, distribution and connectivity. It is essential to generate main assessment quality indicators of regional economic space in accordance with these particular criteria [5].

The system of indicators, characterizing economic space quality of resource-based area in response to technological condition evaluation of industry sectors and allowing to conclude a proper correspondence of the economic situation within a resource-based area between the challenges of economical globalization, has been clarified on the basis of specialists' expert evaluation in the field of regional economy (chart 3).

**Table 1.** Indexes system of economic space qualitative evaluation within a resource-based area.

| Groups of indexes | Economic maintenance of basic indexes |
|---|---|
| **1. Agglomeration and the density of resources distribution** | |
| 1.1. Resource capability indexes | Land resources, natural and inorganic resources, water supply, forest and other natural resources per $km^2$. |
| 1.2. Urbanization and economic density indexes | Quantity of urban settlements, cities of regional and territory significance, agglomerations, megacities and megalopolises in the region. Amount of economically active population, economic density and others. |
| 1.3. Economic enterprise density indexes | Amount of industry, construction activity, farming complex, market, human services enterprises, units of production and non-production infrastructure per $km^2$. |
| 1.4. Communication lines density indexes | Percentage of all the means of transport, communication, internet resources enterprises per $km^2$. The length of railways and highways per $km^2$. |

| | |
|---|---|
| 1.5. GRP amount density indexes | Amount of territory GRP which indicates the way GRP value corresponds to the territory size where the product is manufactured. |
| 1.6. Budget potential indexes | Regional and local taxes and duties, federal taxes allocations in accordance with standards, the capacity of grants, subventions, subsidies, transfer payments and others. |
| **2. Traffic and delivery acceleration, connectivity and disposition** | |
| 2.1. Transport infrastructure development indexes | Rapidity of economic cycles and processes. Capacity of all the means of transport traffic flow, length of highways and railways, federal paths, oil and gas pipeline routes and others means of transport. |
| 2.2. Information infrastructure development indexes | Connectivity and access to Internet host, amount of Internet users, ICT competence, local calls cost, toll line, Internet providers market, telephone lines. Percentage of computers, fixed line and mobile phones users and others. |
| 2.3. Uniformity indexes of regional population distribution (variations) | Variations of typical uniformity distribution of population over the territory of a resource-based region as the standard deviation attitude of the featured interest towards its equidistributed estimate. |
| 2.4. Uniformity indexes of regional enterprises and sectors distribution over the regional territory (variations) | Variations of typical uniformity enterprises and sectors distribution over the territory of a resource-based region as the standard deviation attitude of the featured interest towards its equidistributed estimate. |
| **3. Technological condition of economic sectors** | |
| 3.1.Indexes of sectors with high technological modes | The percentage of sectors with critical technologies, approved by the Russian Federation President's decree № 899, from 7th July 2011. The unit weight of 5th, 6th and higher mode sectors in the economic system. |
| 3.2. Conversion indexes of technological state of economy | Indicators of income transportation sources from low technological mode sales towards the development of technologically high mode sectors (creation of regional reserve funds due to the partial income transfer from mining industry production sales) and others. |
| 3.3.Science and innovation potential | The volume of innovative output. Scientific |

| | |
|---|---|
| indexes<br>Import substitution programms | research charges, the amount of innovative product and processes developments or applications, the structure and the percentage of enterprises involved into innovative activity, number and structure of staff employed in R&D. |
| 3.4. Investment potential indexes | Fixed capital investments due to all the sources of finance, overhaul charges for enterprises and organizations, capital investments into the economic potential development. |
| 3.5. Innovation transfer and extension indexes | The amount of shared innovative projects, created with associated companies. Existence of organizational framework of innovative promotion management. The number of staff employed in innovative promotion management. The volume of innovative promotion charges. |
| 3.6. Indexes of strategic and lawmaking initiatives | Existence of investment projects and business initiatives including those which have been checked by local government, tenders and proposals submitted to alter present legislation and innovative climate in the field of innovative activity. |

The novelty of a modern approach for the qualitative assessment of the resource-based regional economic space is to supplement the system of existing indicators such as “density”, “connectivity” and “distribution” with a fresh block of indicators, defining the technological condition in economic sectors of a resource-based region and innovative potential development of the territory. At the same time it is supposed that the technological condition level in economic sectors of resource-based region depends on primary ratings, which are connected with density, connectivity and distribution, in other words the processes of traffic and delivery acceleration and agglomeration, of economic space quality.

These indicators are apt to evaluate technological condition of regional economy sectors, which should define the tendencies towards the overcoming of raw subordination in regional economies, single out prospective projects of import substitution and develop innovative industries, so that to ensure competitiveness.

On condition that national economy is switching over the innovative course of development modern organizational and qualitative requirements of the economic space within a resource-based region have pointed out the necessity for the extension of basic qualitative assessment criteria inside the economic space, allowing estimation of technological condition among regional resource-based economies that promotes the economical upturn of regional self-sufficiency and their competitive strength as competent market players.

**References.**

[1] Country capital: federal web edition website 2014 Available at: http://www.kapital-rus.ru

[2] Russia in numbers. Brief statistical collection.Rosstat. 2014 p.78

[3] Avramchikova N T, Chuvashova M N 2014 Assessment tools of technological condition of regional resource-based economy sectors *Vesnik. SIBSAU* **1(53)** pp.186-189.

[4] President's Message to the Federal Assembly on 12th December 2013 devoted to the 20th anniversary of Russian Federation Constitution 2013 Available at: http://www.kremlin.ru/news/17118.

[5] Granberg A G 2006 Fundamentals of regional economy *Text-book: State University Higher School of Economics. Moscow* p. 495.

IOP Conf. Series: Materials Science and Engineering **94** (2015) 012018 doi:10.1088/1757-899X/94/1/012018

# BASIC CONCEPTS OF COMPUTERIZED STUDENT-ORIENTED SYSTEM ON LANGUAGE TEACHING

**N V Ivleva**
Senior lecturer of Business Foreign Languages Department
Post-graduate student of System Analysis and Operations Research Department
Reshetnev Siberian State Aerospace University University, Krasnoyarsk, Russia

E-mail: natalie_ivleva@mail.ru

**Abstract:** This article covers the main concepts of a computerized student-oriented system on language teaching which implies more thorough lesson plan design for students who need a strong language command in their professional area. The system analyses all input characteristics and screens lesson plans tailored to individual students which can be integrated into a language course. The designed course is going to be aimed at meeting students' needs as soon as practicable.

Within centuries foreign language mastering is considered to be crucial due to integration and globalization of the world community. More and more people are trying to master a foreign language to use it in their profession field, studies abroad and communication in the international community. However, the area which requires language knowledge most of all is the professional area of a specialist. More and more global companies are establishing their subsidiaries and brunches all over the world that leads to a necessity in learning foreign languages, specifically English, as it is a global language. There is a wide range of methods and approaches in teaching foreign languages; however, there is no computerized student-oriented system on language teaching which could design a lesson plan and a course for specialists in various professional areas.

By a computerized student-oriented system we mean software which could help foreign language teachers design lesson plans and language courses at large tailored to individual students' needs based on personal characteristics, actual language competence and a professional field. The system is supposed to process input data and provide a thoroughly designed lesson plan or a set of plans – a language course – which a teacher could make use of in the educational process. It is expected that the system will refer to a database which will contain a list of language teaching resources for each foreign language taught with certain parameters and choose appropriate ones for screening a lesson plan. This database must be editable to let teachers add new resources and keep it up to date.

In a database we will include the categories of teaching materials according to language skills which are reading, listening, writing and speaking. Depending on students' needs it is essential to focus on some skills more thoughtfully while others could be less stressed on during a language course. It all hinges on the parameters input by a teacher regarding students he/she is preparing to work with. Nowadays, there is a wide range of teaching materials which could be used by teachers to train students in a more effective and encouraging way. Having analyzed a big quantity of resources it is possible to state that all teaching materials can be categorized into coursebooks, grammar and vocabulary books, resources for improving receptive skills (reading, listening) and production skills (writing, speaking). Reading, listening, writing and speaking materials can be found both in coursebooks and separately. Consequently, each category of teaching materials should be described efficiently with due regard to pedagogical and technical aspects of a teaching process.

Another essential role of a system described above is to choose the most suitable method or approach according to the information input into the system. There are 12 well-known methods and approaches which are internationally recognized and used all over the world.

*Grammar-Translation Approach (GTA).* Much traditional language teaching in schools worldwide used to be done in this way, and it is still the predominant classroom method in some cultures. The teacher rarely uses the target language. Students spend a lot of time reading texts, translating them, doing exercises and tests, writing essays. There is relatively little focus on speaking and listening skills.

*Audio-Lingual Method (ALM).* Although based on largely discredited theory, the techniques and activities continue to have a strong influence over many classrooms. It aims to form good habits through students listening to model dialogues with repetition and drilling but with little or no teacher explanation.

*Communicative Approach (CA).* This is perhaps the method or approach that most contemporary teachers would subscribe to, despite that fact that it is widely misunderstood and misapplied. CA is based on beliefs that learners will learn best if they participate in meaningful communication. It may help if we distinguish between a stronger and a weaker vision CA. As a strong aspect students learn by communicating, but they also learn through a wide variety of teaching, exercises, activities and study, with a bias towards speaking and listening work.

*Total Physical Response (TPR).* A method devised by Dr. J. Asher, mainly useful with beginner and lower-level students. Learners listen to instructions from the teacher, understand and do things in response, without being required to speak until they are ready.

*Community Language Learning (CLL).* A method based around use of the learners' first language and with teacher help in mediating. It aims to lower anxiety and allow students to communicate in a more genuine way than is typically possible in classrooms.

*Natural Approach (NA).* Devised by Stephen Krashen, this is a collection of methods and techniques from many sources, all intended to provide the learner with natural comprehensible language so that the learner can pick up language in way similar to a child learning their first language.

*Task-Based Learning (TBL).* A variant of TBL which bases work cycles around the preparation for, doing of, and reflective analysis of tasks that reflect eral life needs and skills.

*Silent Way (SW).* Devised by Caleb Gattegno, this method requires the learner to take active ownership of their language learning and to pay great attention to what they say. Distinctive features include the relevant restraint of the teacher and the use of specially designed wall charts. The use of Cuisenaire rods in mainstream ELT arose from this method.

*Person-Centered Approach (PCA).* Any approach that places learners and their needs at the heart of what is done. Syllabus and working methods will not be decided be the teacher in advance of the course, but agreed between learner and teacher.

*Lexical Approach (LO).* Proposed by Michael Lewis and Jimmie Hill. On the back of new discoveries about how language is really used, especially the importance of lexical chunks in communication, proponents suggest that traditional present-then-practise methods are of little use and propose a methodology based around exposure and experiment.

*Dogme.* Scott Thornbury's proposed back-to-basics approach. Teachers aim to strip their craft of unnecessary technology, materials and aids and get back to the fundamental relationship and interaction of teacher and student in class.

*Content and Language Integrated Learning (CLIL).* This approach refers to teaching subjects such as science, history and geography to students through a foreign language. This can be by the English teacher using cross-curricular content or the subject teacher using English as the language of instruction. Both methods result in the simultaneous learning of content and English.

However, there is one more approach - the eclectic approach - which allows teachers to use techniques and activities from a range of language teaching methods and approaches. The teacher decides which method or approach to use depending on the aims of the lesson and the learners in the group. For example, the class starts with an inductive activity with learners identifying the different uses of synonyms of movement using a reading text. They then practise these using TPR. In another class the input is recycled through TBL, with learners producing the instructions for an exercise manual [1].

Taking all abovementioned methods and approaches into account it is necessary to select those ones which will work well in different programs and with different target audiences. Therefore, to meet students' needs the system should analyze which methods and approaches could be applied to get a required outcome. To do it effectively the system should contain the section with a multiple-choice questionnaire which must be filled in by a teacher upon an interview with new students. With the help of answers provided the system could choose a set of teaching methods or approaches that could suit for working with interviewed students.

However, business world is special comparing to other areas of language teaching. Not all existing teaching methods could work properly with such students. Therefore, the system should involve only those ones which will work in real teaching practice. Based on experience and observation it is possible to state the fact that there is a range of approaches and methods which could be universal for any business student and could be programmed in the database. They are: CA, TBL, PCA and CLIL.

Such methods and approaches are considered to be universal as they help present a foreign language not as a subject to be learnt in isolation from real use but within the context. It stimulates business language learners to master the language as a function for communication, problem-solving and decision-making. It means that with the use of abovementioned methods and approaches business students and students from different professional areas can perform in their fields of speciality and jobs effectively [2].

To satisfy needs of specialists in a wide range of professional areas it could be an efficient approach to design a student-oriented system on language teaching with the focus on personal characteristics, time limits and needs. In the process of research it is planned to develop modified methods of personal characteristics analysis (psychological, physiological, etc.) stressing on the professional area of a learner. According to the main idea of the research all groups of language learners must master a language through multilinguistical needs.

Currently, the research is being carried out at the first stage which involves studying and analysis of various groups of specialists with their personal traits and suitable modes of study. In this article we consider different groups of specialists and define common characteristics for each group.

It is difficult but very crucial to categorize specialists. It is better to group them according to types of sciences. The system of numerous and multivarious researches distinguished by an object, subject, method, fundamentality degree, sphere of application, etc. prevents from creating a common classification of all sciences and, consequently, professions. It is possible to categorize all sciences only in general terms. They can be natural, technical, social sciences and humanities.

Natural sciences include sciences about space, its organization and development; sciences closely connected with our planet such as geology, geophysics and geochemistry; physical, chemical and biological sciences; and also studies on human being development, his/her origin and evolution. Technical sciences are mainly based on natural sciences. In technical sciences a great impact is made on development of technologies (radio engineering, thermal technologies, etc.). Social sciences study societies, their development and attitude to life (sociology, economics, etc.). The last group but not the least one is humanities. All humanities are sciences about the spiritual and intellectual world of a person, about his attitude to the environment, society and other people around. There are also such sciences which can be related to different groups such as medicine, ecology, engineering psychology, etc. Philosophy, mathematics, informatics, cybernetics play an important role for all groups of sciences and are applied in all types of research.

All university degrees and professions stem from those sciences. Nowadays, we have a big range of specialists having professions related to one scientific area or some of them. It all depend on duties and functions at work. According to the survey, performed among people, who have to reach a certain level of language fluency, there are certain professions which require language knowledge. Such professional areas are information technologies, trade, construction development, management, engineering, medicine, tourism, translation/interpretation, teaching, aviation, advertising and PR, logistics, journalism, marketing, entrepreneurship, law, copyright. All these professions are originally related to technical, social sciences, humanities and, obviously, they are in need of English at different angles and levels.

All categories of specialists have absolutely different duties and responsibilities. Technical specialists mostly work with machines, facilities and other engineering installations which they program, control and fix errors occurred. They use their analytical skills and try to foresee problems in the process. While social workers and people working for humanity areas mostly deal with people and servicing them.

Regarding language needs technical specialists are more focused on reading skills. In their professional areas they read a wide range of technical literature, for instance, manuals, guides, instructions and texts dedicated to technical requirements and descriptions. The main problem for such specialists is a lack of professional literature in their mother tongue. To be up-to-date and aware of all modern tendencies and trends they have to obtain information through English resources, which are developed and published by most countries leading in those areas in English to popularize their researches and developments globally.

Specialists whose professions originally come from social sciences are eager to master the language to be able not only to read professional texts developing their reading skills, but also they pay much attention to their writing skills as they work with a big amount of information, statistical data and they have to summarize all of it in a report or any other document which provides data to a target audience (management, society, etc.).

The last group of specialists who intensively study English is humanities specialists. Due to their professional specifics they communicate a lot, give presentations, negotiate and speculate. All of these functions give an idea that they learn the language stressing on speaking skills. They attend conversational lessons, desire to boost their language through transferring their passive vocabulary into active one and wish to be as fluent as possible because their English fluency influences their workout and result of communication.

In conclusion it is important to state that the computerized student-oriented system on language teaching could play a crucial role in improving the language teaching process and would help students master the language in a more efficient way. The system will represent software which will have a teacher-friendly interface. The only this a foreign language teacher should do is to put in data required by the system and get an automatically-designed lesson plan and the whole course with the aim, objectives, competences and other information thanks to which it will be clear for both for a teacher and student what results will be achieved after course completion.

## References

[1] Jim Scrivener. Learning Teaching. The Essential Guide to English Language Teaching, 3rd edition, Macmillan, 2011. – P. 418
[2] Sino-US English Teaching, ISSN 1539 – 8072, February 2012, Vol. 9, No. 2, 913 – 917.

# DANGEROUS FAILURES IN MULTIFUNCTIONAL SYSTEMS*

**P A Kuznetsov, I V Kovalev and P V Zelenkov**
Siberian State Aerospace University named after Academician M.F. Reshetnev
31 "KrasnoyarskiyRabochiy" prospect, Krasnoyarsk, 660037, Russia.

E-mail: forubox@yandex.ru

**Abstract.**The paper describes the peculiarities of multiple indicators of reliability of automated control systems. Along with the dependability considered an indicator such as the dangers of system failure. Showing particular account of the reliability of the automated control systems when considering their versatility. We study methods of lowering the probability of dangerous failure. Such method could be the reallocation of resources needed for inclusion of backup elements from module carrying a low danger to modules, carrying high-dangerous failures. Another method is identification of least important functions and transfer resources, needed for inclusion of backup from them to most important and dangerous modules, and preventing threats by blocking modules.

## I Introduction

This paper describes the peculiarities of analysis of indicators of reliability for automated control systems (ACS). Reliability of technical systems is one of the most important indicators.

Each ACS includes variety of different elements, executing various functions. Moreover, one element may perform several functions, and, conversely, a single function may be performed by several elements.[1,2]

Reliability of the system is its possibility to keep all of the parameter, qualifying possibility of executing all of the functions in determined range in process of time [3].

There is a variety of different reliability-related indices, for example fault-tolerance, safety, efficiency etc. [4, 5]. In a modern reliability theory fault-tolerance is traditionally accepted as primary reliability parameter. Fault-tolerance usually means the amount of failed elements at the same time which would lead to the failure of the system as a whole. The more elements that are required for the failure of a system, the more fault-tolerant this system is [6].

One of the ways to improve it is redundancy. In redundant systems elements categorized by a role to main and backup. Main elements are the elements of the object, which failure in a case of absence of backup cause failure of the system. Backup elements are the elements purposed to keep system intact in a case of main element failure [7].

## II Methods of infallibility increase

* The research was done with the financial support of the Ministry of Education and Science of the Russian Federation in accordance with the agreement № 14.574.21.0041, unique identifier RFMEFI57414X0041

There is various method of composing reliable systems, based on stochastic or determined principles. Main idea of those methods is solving the optimization problem - maximization of the objective function subject to constraints.

Reliability indices presented by a list of values. Infallibility is the probability of lack of failure for a certain period of time.

A typical objective function is a function of entire system infallibility, which is expressed as the product of all modules infallibility.

$$P = \prod_{i=1}^{N} P_i$$

$P_i$ – probability of faultless functioning of module i

N – number of modules.

Infallible structure is composed by iterative procedure of adding in system modules with lowest infallibility rate [8].

Function, computing the priority of reservation

$$Ra_i = \frac{1}{P_i}$$

**III Alternative reliability indices**

But the conception of maximization of infallibility has certain disadvantages. It should be considered that with any probability of failure-free operation remains opposite probability - the probability of failure. Even if arbitrarily large incensement of reliability, we can not rule out system failure, as this process is probabilistic. [9,10]

Also, any ACS is composed by physical modules, operating physical matter and energy of technological processes. And in a case of failure correct operation of matter and energy interrupts. There is an emergency situation that may be dangerous for the personnel and infrastructure.

Dangerous failures could be appear in modules. System, composed by nodules, having high possibility of absence of failure, anyway could be highly dangerous. Because of that we need to purposely lowering possibility of failure of highly dangerous modules.

To operating the conception of danger we need to have quantity measure of it. For example, there is a measure by class of danger of chemicals in a case of chemical dangers and by energy for energetic[11].

**IV Redistribution of resources for backup**

Considering dangerous failures, it is necessary to further increase the probability of failure-free operation for the modules highly hazardous failures. One way is to redistribute resources spent on redundant modules with less dangerous modules to more dangerous. We should change the priority of the backup priority function.

$$Ra_i = \frac{D_i}{P_i}$$

Di is danger of failure of module i.

But even the simple redistribution is lowering the possibility of dangerous failure, at the same time it could increase possibility of system failure.

Possible way to solve this dilemma is to change the objective function of the system.

Automated control systems are multifunctional. ASC generally have one main purpose, which could be divided into sub-tasks, performed by certain functions of the system. Each function plays determined role in completing of the system purpose. Some of the functions take more, and some - less participation in the system. The failure of an individual component leads to the inoperability of that component and potentially other components, but it does not lead to the inoperability of the whole system. [12, 13].

Thus, we can, for the sake of improving the safety of the system to announce some functions less important and to redistribute resources from the modules of these functions into modules with dangerous failures. We should determine which sequence of elements is performs which functional task. This procedure of the separation of the existing system into subsystems, components, called decomposition. Decomposition as dissection process allows considering any system being studied as a complex consisting of separate interconnected subsystems, which in turn, can be subdivided into parts. In decomposition each dismemberment has its own level.

Base system is located on level zero. After its decomposition we get first level systems. Decomposition of them results second level subsystems etc.

After functional decomposition we should to compute and appoint importance of the modules executing functions. Total probability of failure could be raised, but probability of main function of the system failure will not be affected. Also, probability of dangerous failure will be lowered because of the redistribution of resources from less important functions.

## V Blocking modules

But in addition to reducing the probability of failure there are direct ways to reduce the danger caused by failure - the inclusion of specific modules, blocking harmful effects before they impact on staff or infrastructure [14]. Such modules may simply discourage dangerous influences to cause harm, and actively eliminate them. An example of blocking modules, preventing the effects is individual and collective protective equipment.

Protective equipment - the means used by workers to prevent or reduce the impact of harmful and hazardous production factors, depending on the purpose protective equipment is divided into classes:

Means of normalization ambient air of industrial premises and workplaces, means of protection of high-level ionizing radiation, means of protection from high electric and magnetic fields etc [15].

And the example of the danger of liquidating modules can be modules neutralizing emergency chemically hazardous substances. The neutralization of poisonous substances - action aimed at the destruction of the toxic properties of substances, based on the chemical conversion of poisonous substances into non-toxic products

The need to include these modules is determined by the safety requirements, and they includes before the redundant functional modules.

## Conclusion

Thus, we conclude that the need for the construction of automated control systems complement the principles of redundancy by other principles. This will take into account the full reliability indices.It is necessary to separate the modules which make up the system of importance. This allows you to optimally allocate resources to reserve the most important and dangerous failure of modules.You must also take into account the flow of resources into units to ensure the safety of the system. Review of the operation of these modules in the design phase allows you to specify the consumption of resources on the system and its safety.

**References.**

[1] GOST 24.701-86 Edinaya sistema standartov avtomatizirovannykh sistem upravleniya. Nadezhnost' avtomatizirovannykh sistem upravleniya. Osnovnye polozheniya *Unified system of computer control systems standards. The reliability of automated control systems. Fundamentals* (In Russ.) Moscow. Standards publ.. 11 p.

[2] Kuznetsov P A, Beschastnaya N A, Bakhmareva K K, Antamoshkin O A, Antamoshkin A N 2012 Modification of the Volkovich`s-Michalevi`s method to optimize costs in the synthesis of fault-tolerant information systems *Vestnik SibGAU* **6(46)** pp. 97-100 (In Russ.)

[3] Kurochkin Yu A, Smirnov A S, Stepanov V A 1993 Nadezhnost' i diagnostirovanie tsifrovykh ustroystv i sistem *Reliability and diagnostics of digital systems* (St.Petersburg. St.Petersburg University Publ.) 320 p.

[4] Kovalev I V et al 2015 Model of the reliability analysis of the distributed computer systems with architecture "client-server" *IOP Conf. Ser.: Mater. Sci. Eng.* **70** 012009 doi:10.1088/1757-899X/70/1/012009

[5] Kovalev I et al 2015 The efficiency analysis of the automated plants *IOP Conf. Ser.: Mater. Sci. Eng.* **70** 012007 doi:10.1088/1757-899X/70/1/012007

[6] Kovalev I V et al 2015 The control of developing a structure of a catastrophe-resistant system of information processing and control *IOP Conf. Ser.: Mater. Sci. Eng.* **70** 012008 doi:10.1088/1757-899X/70/1/012008

[7] Kucher V Ya Osnovy tekhnicheskoy diagnostiki i teorii nadezhnosti: Pis'mennye lektsii *Fundamentals of technical diagnostics and reliability theory: Lecture* (In Russ.) Available at: http://window.edu.ru/resource/967/24967 (accessed 3.05.2015)

[8] Sugak E V, Vasilenko N V 2000 Nadezhnost' tekhnicheskikh sistem. Uchebnoe posobie dlya vysshikh uchebnykh zavedeniy *Reliability of technical systems. Study guide* (Krasnoyarsk. NII SUVPT Publ.) 608 p.

[9] Ryabinin I A 2007 Nadezhnost' i bezopasnost' strukturno-slozhnykh sistem *Reliability and safety of structural complex systems* (St.Petersburg. St.Petersburg University Publ.) 276 p.

[10] Okhtilev M Yu, Sokolov B V 2006 Intellektual'nye tekhnologii monitoringa i upravleniya strukturnoy dinamikoy slozhnykh tekhnicheskikh ob"ektov *Intelligent technologies for monitoring and management of the structural dynamics of complex technical objects* (Moscow, Science Publ.) 410p.

[11] Akimov V A, Lapin V L, Popov V M, Puchkov V A, Tomakov V I, Faleev M I 2002 Nadezhnost' tekhnicheskikh sistem i tekhnogennyy risk. Uchebnoe posobie *Reliability of technical systems and technogenic risk* (Moscow. Delovoy ekspress publ.) 368 p.

[12] Kuznetsov P A 2015 Dependent failures in multifunctional systems *Vestnik SibGAU* **1(57)** pp. 86-91.

[13] Kovalev I et al 2015 The minimization of inter-module interface for the achievement of reliability of multi-version software *IOP Conf. Ser.: Mater. Sci. Eng.* **70** 012006 doi:10.1088/1757-899X/70/1/012006

[14] Kovalev I V, Kuznetsov P A, Zelenkov P V, Shaydurov V V, Bakhmareva K K 2013 To the question of the reliability of automated control systems with blocking protection modules *Pribory* **6** pp.20-24 (In Russ.)

[15] Rossiyskaya entsiklopediya po okhrane truda *Russian encyclopedia of labor protection* (Moscow. NC ENAS Publ.) 2007

IOP Conf. Series: Materials Science and Engineering **94** (2015) 012020 doi:10.1088/1757-899X/94/1/012020

# FAULT INJECTION VIA ON-CHIP DEBUGGING IN THE INTERNAL MEMORY OF SYSTEMS-ON-CHIP PROCESSOR*

**S.A. Chekmarev[1], V.Kh. Khanov[2]**
Siberian State Aerospace University named after Academician M.F. Reshetnev (SibSAU)
31 “Krasnoyarskiy Rabochiy” prospect, Krasnoyarsk, 660037, Russia.
[1] Engineers of SibSAU
[2] Ph.D., assistant professor of SibSAU

E-mail: khvkh@mail.ru

**Abstract.** The paper presents an on-chip debugging method for the injection of single faults in the processor cores of systems-on-chip. The method consists in the placement of faults injection infrastructure in a system-on-chip as an intellectual property core. This simplifies the fault injection environment, reduces delays injection and improves the performance, as well as allows doing long autonomous campaign for injection of faults without the use of external devices.

## 1. Introduction

Fault Tolerance (FT) is an important requirement for digital equipment working in hard conditions, such as in space. Today the majority of such equipment is based on microprocessors. The greatest danger for microprocessors working as a part of equipment systems on board spacecraft constitute Single Event Upset (SEU), caused by cosmic ionizing radiation. These events can change the logical values in the elements of the internal memory of the microprocessor, such as registers and cache [1]. SEU is the most frequent cause of faults, so efforts for the FT design of architecture microprocessors mainly focus on protection from these events.

Fault Injection (FI) is a widely accepted solution to measure the SEU sensitivity of microprocessors and other digital devices. There are several different FI solutions which have their benefits as well as disadvantages. Some solutions are universal; others are dependent on the capabilities of the System Under Test (SUT). The majority of modern microprocessors include circuitry On-Chip Debugger (OCD), which allows access to internal resources. This feature can be used to modify the registers and memory, providing a useful mechanism for FI.

Existing approaches to inject faults using OCD [2, 4, 5, 6, 8] have essential limitation: most of them use host PC and external to the SUT debugger associated with host PC for the FI campaigns, i.e. FI source is out of the SUT. This solution complicates the fault injection environment and limits the performance of FI campaigns. However microprocessors type system-on-chip (SoC) may contain the necessary infrastructure for the FI, so the injection will be generated inside the microprocessor. This simplifies the FI-system and reduces the time for performing a single injection and increases the productivity of FI-companies.

This paper presents a new fault injection approach to inject faults in SoC-microprocessor that supports an OCD. Hardware module FI which autonomously produces FI experiments without external devices is embedded in the SoC-microprocessor as an IP-core. This simplifies the implementation of fault injection environment, avoids the communication between the target system and the software debugging tool which operates in the host PC. The feasibility of the proposed mechanism has been proved injecting faults on a LEON3 processor. The analysis of the results shows that this solution is an efficient, fast and little intrusive method to evaluate SEU sensitivity in SoC-processors.

* The research was done with the financial support of the Ministry of Education and Science of the Russian Federation in accordance with the agreement № 14.574.21.0041, unique identifier RFMEFI57414X0041

## 2. Fault Injection methodologies via on-chip debugging

Embedded debugging capabilities, included in most current microprocessors, allow the access to their internal resources for reading and writing. This feature can be used for fault injection. Injecting a fault via a circuitry OCD involves the following steps: setting a breakpoint via the circuitry OCD and waiting for the program to reach the breakpoint; reading the value of the target location (a register or a memory word) via the circuitry OCD; manipulating this value and then writing a new, faulty value back to the target location; resuming the program execution via a command sent to the circuitry OCD.

In most cases, OCD -approach uses the external debugger related with OCD a JTAG, Nexus or BDM interfaces. In [2] the fault injection environment architecture using the standard IEEE JTAG 1149.1 [3] is presented. It consists of a computer, a specialized FI device and SUT having circuitry OCD. The FI device includes a fault injection controller, a fault list, and golden results, debug commands module and JTAG controller. This solution allows carrying out the functions of controlling injections, preparation of the planned events and storage of results of fault injections not in the external host computer but with the help of the specialized FI device. The advantage is the ability to conduct long-term autonomous multiple FI experiments without a host computer, for example, a thousand or more events. The disadvantage of this architecture, like the previous one, is the complexity of the FI campaign, as it requires an external specialized FI device.

In [4, 5, 6] presents several FI-system using a standardized approach to debugging NEXUS [7]. The microprocessor of SUT contains circuitry OCD compliant with Nexus. Nexus standard defines the requirements for the debugging device and the interface that connects the debugging device and circuitry OCD. The disadvantage of the system is that the debugging device compatible with the standard Nexus is not designed to make faults. Is why it is equipped with additional modules for FI. In another example, a part of the expanded functional typical for [4], namely, FI controller, is located directly in the circuitry OCD, which increases productivity. The disadvantage of these infrastructures FI is the complexity, because it includes an external debugging device. In addition, the last solution has high intrusive, which reduces its practical significance.

In [7] simulated SEU error injection via OCD in the internal memory of the processor LEON3 [8], which is a SoC, is presented. OCD of LEON 3, in this case called Debug Support Unit (DSU), has its own direct interface with an on-chip bus, and, consequently, an external host PC has direct access to it. To improve the performance of FI the host PC is connected to the SUT via a high speed channel, for example, JTAG or Ethernet. FI campaigns are fully managed by software of a host PC, which is a disadvantage.

The objective of the work is to develop a simple and productive infrastructure FI, which allows carrying out long-term stand-alone experiments to test fault tolerant microprocessors such as system on chip to single faults via OCD-method.

## 2. Proposed solution

The goal is achieved by integrating FI infrastructure in SoC-microprocessor system as a specialized IP-core. This infrastructure includes Fault Injector, containing Fault Injection Controller, Fault Generator Drive and Golden Results. The environment architecture is presented in Fig. 1 and is intended for FPGA implementation.

The process of faults injection in the internal memory of the microprocessor core occurs as follows. Host PC initiates the start of the experiment by injection into the internal memory of the processor core. Fault Generator determines the time, the address and the position of the internal memory cell for making faults. This action can be performed automatically by a uniform random law or can be read from Fault List, which is pre-loaded into the internal memory of the FPGA from host PC. Fault Injection Controller, using the information received from the Fault Generator, via OCD writes an error to the required bit and the desired address of internal memory by way of inverting its content. To make a fault the OCD stops the processor core, reads the desired internal memory cell, inverts the required bit and writes the result back. Then Fault Injection Controller and OCD, interacting by an on-chip bus, repeat the same steps similarly, when Fault generator determines the next injection. If the mechanism of error detection and correction detects a fault during the performing of the current running program by the processor core of internal memory, then depending on the result the signs «fault is detected and corrected» or «fault is detected and not corrected» are saved in internal memory status register. Simultaneously the Fault Injection

Controller polls the internal memory status register for finding facts of faults detection. To do this, it gives appropriate instructions to OCD, which reads the status register of on-chip memory, and analyzes the impact of faults. If signs of the impact of faults are established, this information is recorded in the Golden Results. FI process continues until the end of Fault List, which has been pre-loaded, or until the host PC forcibly stops the Fault Generator and thereby stops FI-campaign.

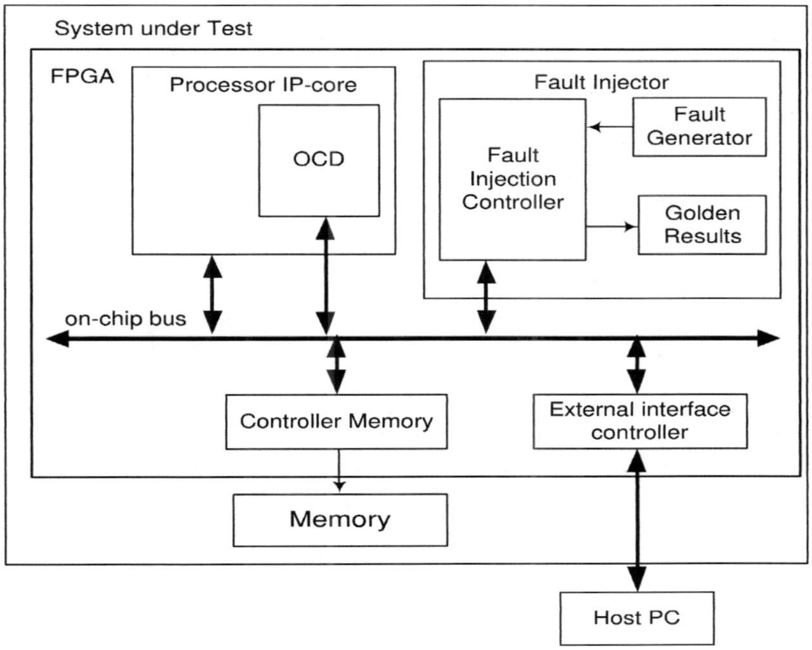


**Figure 1.** Fault injection environment architecture.

Upon completion of the experiment the results of injection of Golden Results are transferred to the host PC. If the injection is performed automatically by the Fault Generator, then the information on the number of faults introduced during the experiment is also transmitted to the host PC. The results obtained are processed in the host PC to assess the effectiveness of the mechanism of error detection and correction of internal memory of the processor.

The proposed solution allows for long-term experiments on the injection of faults of almost any number (hundreds, thousands, tens of thousands, hundreds of thousands events) due to the use of a Faults Generator. It allows for the realization of not only the random uniform law of injection, but also any other law, including that simulating the degradation of the system under test during its operation due to accumulation of doses of radiation from outer space.

Fault Injector is an independent IP-core not affecting the remaining parts of the microprocessor system. Therefore, after all the experiments Fault Injector can be removed, and thus a complete microprocessor system will not have it. If Fault Injector is left as a part of a system composed of microprocessors, it can be used to diagnose system fault tolerant in microprocessor system exploitation as, for example, an onboard computer of a spacecraft.

### 3. Experimental results

The effectiveness and applicability of the proposed approach was evaluated in an open processor LEON3, equipped with a fault tolerant mechanism based on Hsiao code [9]. The main experimental studies were carried out on the board of a processor module [10] based on A3PE3000 FPGA of MicroSemi company [11], developed for small spacecraft TabletSat-Avrora [12]. Implementation of the system of FI for LEON3 is presented in Fig. 2. OSD in LEON3 called DSU (Debug Support Unit) as previously noted. DSU as well as all the IP-core has access to the on-chip bus AMBA, but also directly related to the core processor. DSU described above introduces faults in registers and cache. Registers and caches are EDAC-module for error correction, and status registers to store the results of the current FI. Fault Injector refers to the processor core through the AMBA, and thereby ensured its weak connection with it and, therefore low intrusiveness. After finishing the all FI-companies needed to test a processor the Fault Injector is removed from the system leaving no trace of it.

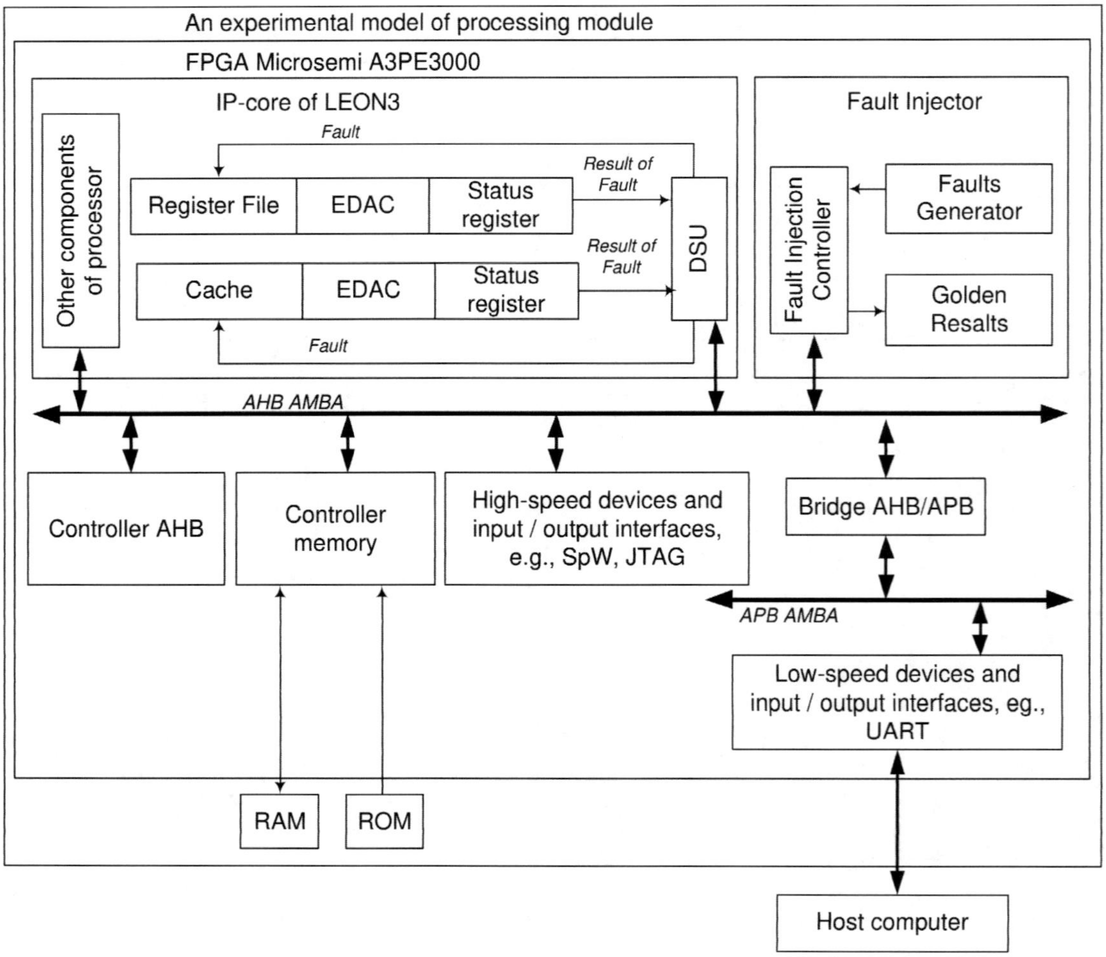


**Figure 2.** Implementation of the system of FI for LEON3.

FI campaigns were carried out with the help of automatically generated random uniform sequence of injections and injection of pre-formed in the host PC and loaded into FPGA memory of fault list separately for the register file and for the processor cache. Ongoing campaigns limited total number of events (100, 500, 1000 and 2000) for limiting the intensity of faults 1, 10, 100 or more events per minute. Thus, to confirm the functionality and feasibility of the proposed approach has been a large number of experiments. The article provides only part of the results.

Table 1 shows the results of injection in the ALU register file of the processor. Fault Generator injection produced by a random uniform law. During the tests, the processor performing the test program that in the loop consistently appealed to the registers with read/ write operations. In the first register recorded value of 0x11111111. Next is his reading. When mismatch of the result is displayed error information. Then, in the next register writes the value of the previous register plus 0x11111111 and then the process was repeated.

**TABLE 1**. Injection results in the register file

| Number of injections | Number of corrected failures in the register file of processor. | | |
|---|---|---|---|
| | 1 event per min | 10 event per min | 100 event per min |
| 100 | 45 (45%) | 51 (51%) | 58 (58%) |
| 500 | 263 (52,6%) | 279 (55.8%) | 218 (43.6%) |
| 1000 | 581 (58,1)% | 613 (61.3%) | 641 (64.1%) |
| 2000 | 1118 (55,9%) | 1097 (54.8%) | 1358 (67.9%) |

Presented in Table 2 the results of injection into the cache of the processor, which is also, produced via of automatically generated random sequence uniform injections. The maximum intensity of failures during the experiment was 100 events per minute. Used the program « RTEMS -Tasks», part of the example programs for the processor LEON3, running under the operating system RTEMS. This program demonstrates the basic features of switching between tasks in RTEMS. Three tasks are generated that are executed in parallel. Each of them makes requests to the system timer and calculates the current system time in the format <*h: min: sec month: day: year*>. The value of the system time is displayed to the user on the console. Successful operation of this program for a long time under the influence of single injection into the cache memory of the processor confirm proper functioning of the functions of the fault tolerant.

**TABLE 2**. Results injections into the processor cache

| Number of injections | The processor cache | | | |
|---|---|---|---|---|
| | instruction cache | | data cache | |
| | Tag | Data o | Tag | Data |
| 100 | 62 (62.0%) | 57 (57.0%) | 48 (48.0%) | 54.0 (54%) |
| 500 | 279 (55.8%) | 218 (43.6%) | 261 (52.2%) | 227 (45.4%) |
| 1000 | 529 (52.9%) | 513 (51.3%) | 678 (67.8%) | 612 (61.2%) |
| 2000 | 1283 (64.5%) | 1336 (66.8%) | 1227 (61.3%) | 1340 (67.0%) |

In the present case detection of single faults has been corrected fault tolerance mechanism. However, not all injections were discovered - the percentage of detected faults is indicated in parentheses. This is due to that some of them occurred in the memory area, which did not contain actual data at the time of entering of failure or data were subsequently overwritten of processor, before he turned to these areas with the read operation. Thus, the experimental results depend on the generated fault and on the specifics of the software testing.

The evaluation of the developed architecture performance was made in comparison with the approach presented in [8], when the injection performing is fully guaranteed by the host PC software. The same configuration of LEON3 processor with the same EDAC on the basis of Hsiao code and the same experimental conditions were used for the injection performing.

Analysis of injection procedures in the internal memory of the processor LEON3 showed that the delay at the time of injection is.

$$t_d = \frac{N_0 \pm N_b}{F},$$

were $N_0 = 54$ – the number of cycles required to make one injections, $N_b = 7$ – the maximum number of clock cycles required for the release of the bus AMBA, when her employment by another device, $F$ – system frequency, i.e. $54 \pm 7$ system clock cycles. When the system frequency is 25 MHz, the delay will be $2160 \pm 280$ ns. Floating value of the time delay associated with the feature of on-chip bus (at the start of injection the bus may be occupied by another master, which is part of a system-on-chip). However, you can ensure that the injection is performed for 61 clock cycle, allowing you to accurately predict the moments of the injection. Thus, compared with the approach shown [8] when the injection is completely provided software of host PC, proposed method allows to control the course of the FI experiment up to 280 nsec. This is achieved due to the lack of dependence on the host PC software and there is no need to exchange information on the external interface. These arguments allow us to classify proposed method of fault injection as real-time.

To evaluate the redundancy considered architecture has been implemented for a different number of faults and for FPGA produced by the companies Microsemi and ALTERA. The results presented in Table 3 show that the proposed approach is not resource-consuming. With the increasing size of the list of events the number of logic cells necessary for the implementation of the method remained virtually unchanged. The need for on-chip memory for multiple programmable and single programmable flash-FPGA of Microsemi company increases faster. This is primarily due to the characteristics of FPGA data. However, the application of the function of FI automatic generation in the proposed method allows reducing the required amount of on-chip memory FPGA. Fault Generator can be automatically initialized for entering new portion of injections. Thereby the need for memory blocks reduces.

**TABLE 3.** Results of the synthesis of a faults injector in FPGAs

| Size of Fault List | Microsemi A3PE3000L | | Microsemi RTAX 1000S | | ALTERA Cyclone IV | |
|---|---|---|---|---|---|---|
| | Cells | FFs | Cells | FFs | LEs | FFs |
| 100 | 1285 (1.7%) | 2 (1.7%) | 301 (1.5%) | 2 (5%) | 370 (<1%) | 2.176 (<1%) |
| 500 | 1301 (1.7%) | 3 (2.6%) | 386 (2.1%) | 3 (8%) | 369 (<1%) | 8 (<1%) |
| 2000 | 1337 (1.7%) | 9 (8.0%) | 448 (2.4%) | 9 (25%) | 379 (<1%) | 34 (<1%) |

## 4. Conclusion

This paper presents a new fault injection method aimed at measurement of SEU sensitivity in SoC-microprocessors. The proposed solution is to integrate FI-infrastructure in SoC-microprocessors as a specialized IP-core. This enables to make FI-campaigns which do not need the debugging software completely autonomous from the host PC that in its turn makes the fault injection process faster. This solution is suitable for Soc-microprocessors in which OCD circuitry has its own direct interface with the on-chip bus. IP-core of FI-infrastructure is removed from the SUT at the end of the debugging.

Experimental results on an LEON3 processor have been presented. They illustrate such capabilities of the approach as long-term experiments conducted with any number of failures, small time delay of injection, and low resources consumption of SUT.

## References

1. S. Mukherjee, "Architecture Design for Soft Errors", Elsevier, 2008 – 337 p.
2. M. Portela-Garcia, C. Lopez-Ongil, M. Garcia-Valderas, L. Entrena, "A Rapid Fault Injection Approach for Measuring SEU Sensitivity in Complex Processors", IOLTS 2007, 13th IEEE International On-Line Testing Symposium – Heraklion, Crete, Greece – July 8 to July 11, 2007– pp.101-106.
3. "IEEE Standard Test Access Port and Bondary-Scan Architecture", IEEE Std 1149.1, 2001 – 212p.
4. A. Fidalgo, G. Alves, J. Ferreira, OCD-FI: on-chip debug and fault injection, in: International Conference on Dependable Systems and Networks, June 2006 – pp. 214-219.
5. A. Fidalgo, J. Ferreira, "Using NEXUS Compliant Debbuggers for Real Time Fault Injection on Microprocessors", SBCCI'06, 19th Symposium on Integrated Circuits and System Design, Minas Gerais, Brazil – August 28 - September 01, 2006 – pp. 214-219.
6. A. Fidalgo, M. Gerigota, G. Alves, J. Ferreira. "Real-time fault injection using enhanced on-chip debug infrastructures", Microprocessors and Microsystems № 25, 2011– pp. 441-452
7. "The Nexus 5001 Forum Standard for a Global Embedded Processor Interface version 2.0", IEEE-ISTO 5001, 2003 – 166p.
8. J. Gaisler, "LEON3-FT-RTAX SEU test results Gaisler Research", Issue 1, December 7, 2005 – 8p.
9. "SPARC V8 32-bit Processor LEON3/ LEON3-FT Companion Core Data Sheet Template Design", Gaisler Research, 2010 – 41 p.
10. M. Y. Hsiao, "A class of optimal minimum odd-weight column SEC-DED codes", IBM J. Res. Develop., vol. 14, no. 4, Jul. 1970 – pp. 395-401.
11. V.Kh. Khanov, S.A. Chekmarev, V.A. Shakhmatov, M.Y. Vergasov, F.A. Lukin, "Development of equipment information exchange system of onboard control complex small spacecraft" ["Razrabotka apparatury sistemy informacionnogo obmena malogo kosmicheskogo apparata"], Vestnik SibGAU. no. 5 (49), 2013, pp. 149-153 (In Russ.).
12. "ProASIC3 FPGA Overview", available at: http://www.microsemi.com/products/fpga-soc/fpga/proasic3-overview, (last visited: January 2014).
13. "Sputnix", available at: http://www.sputnix.ru/ru, (last visited: March 2014).

# REPRESENTATION OF THE AWARENESS REFLEXIVE STRUCTURE WITH REGARD TO THE CONFLICT INVOLVING SIGNATURES*

**M Styugin**
Research Fellow, Siberian State Aerospace University Named After Academician M.F. Reshetnev, Krasnoyarsk, Russia

E-mail: styugin@gmail.com

**Abstract.** This paper offers a method for analysis of conflict situations when utility of agents is replaced by rules of decision making. This approach simplifies review of a conflict situation, though it imposes significant restrictions to conflict conditions. A conflict shall be presented as separate transitions and Boolean readiness of agents to perform those transitions. These restrictions can be seen when a model is formulated. This approach appeared in modeling of conflict situations in security systems and it corresponds to analysis of intruder-defender systems the most. Models provided in the present paper enable to calculate an object's readiness to take active actions depending on the condition of its awareness structure. Awareness structure includes the view on indicators of readiness to take active actions by other participants of a conflict and their view on the indicators of the others. A notion of a system researcher is introduced based on the model. The system researcher creates its own awareness structure upon researching structures of the other participants. Conditions for a researching task for two awareness structure researchers which analyze structures of each other are provided herein and it is ascertained which of the participants and in which case is able to solve the task. The paper provides examples of applying mathematical models in real life situations.

## 1. Introduction

The paper offers a specific method of analysis in a theoretic game situation, when the utility function is not used explicitly. In other words, the analysis of a game situation with specific features is observed herein. The difference of the signature model from the classic theoretic game situation task is only in representation form. Actions of players depend on the reflection level. That dependence was described by V.A. Lefebvre in reflexive research and introduced as a logical element in reflexive games [1, 2]. Later on D.A. Nivikov and A.G. Chkhartashvili named that strategic reflection [3] and reflexive games in addition to strategic reflection include informational reflection, that is values of different elements of the awareness structure. This particular definition of reflexive games is used in the present paper.

We can see a dependency of a player's activity throughout the whole awareness structure by replacing the utility function with rules of decision making. In some cases it is more

* The research was done with the financial support of the Ministry of Education and Science of the Russian Federation in accordance with the agreement № 14.574.21.0126, unique identifier RFMEFI57414X0126

convenient than introduction of a utility function in applied tasks as we can easily find the whole set of awareness structures in the conflict favorable for us.
The signature model, which is presented in this article, evolved out of conflict situations. It mostly corresponds to the analysis of intruder-defender systems. Clearly, the classic theoretical game model is widely considered for these type of systems. Let us consider the following articles for example [4-6]. Regardless that they used a different method of the game theory in these works, all of them used the qualitative method based on the utility function. It's not always possible for the researcher to get the initial value of the utility function. The qualitative characteristic is more convenient in order to answer the question of whether the agent benefits from this transaction or not. This qualitative characteristic allows us to build a Boolean expression that lets us make decisions.

## 2. Model Formulation

Not every utility function can be represented as a Boolean signature, but only with two possible actions for each player. That means that a conflict situation has to be reduced to separate transitions, when each of them is related to the activity of one of the players (take actions or stay inactive). Here is an example of such model.
Suppose there are two agents in the system defined as *x* and *y*. Agent *x* wants to take some actions (they may be interpreted as attack methods), and agent *y* needs to take a counteraction (protection methods) against each action of *x*. Hence, a conditional payoff matrix for the agents that has the form as shown in fig.1

$x_+ =$

| –\\+ | 0 | 1 |
|---|---|---|
| 0 | 0 | +1 |
| 1 | 0 | -1 |

$y_- =$

| –\\+ | 0 | 1 |
|---|---|---|
| 0 | 0 | -2 |
| 1 | -1 | -1 |

**Figure 1**. The payoff matrix of agents.

Columns marked with "+" denote actions of the agent trying to make a transition in the system. In that case it is agent *x*. Lines marked as "– " denote actions of agent *y* that resists the transition in the system.
1 stands for agent taking actions, 0 stands for inactivity.
According to the first matrix when the agent does not take actions (attack) then it gets no payoff. When it takes actions and the counterpart does not oppose then its payoff is +1, when the counterpart opposes to the transition then the payoff is -1 (attack failed + cost of taking actions).
According to the second matrix when the agent takes actions then its payoff is always -1 (costs). When it does not take actions and the counterpart does not take actions, then the respective payoff is 0, when the counterpart acts (attacks), then the damage is -2.
Agents are aware of the actual matrix values of each other. And resistance to transition in the system is always preventive, which is quite reasonable in terms of security systems.
The game above does not have a solution in pure strategies, and thus agents have to be governed by a guaranteeing strategy. In this case guaranteeing strategies are based on the reflection level. Decision of each agent on the choice to act or stay inactive is based on the length of the chain of considerations "I think that he thinks of what I think...".

IOP Conf. Series: Materials Science and Engineering **94** (2015) 012021 doi:10.1088/1757-899X/94/1/012021

Let us unify notation of such strategies with composing clearer conclusion principles of agents. Therefore 0 and 1 shall denote agent's readiness to take a specific action (counteraction) in the example provided. The initial state of the object in a conflict shall be denoted as $O$ and suppose that the object can be transferred in to the $O'$ state as a result of actions taken by the subject(s). Such transition is possible after actions taken by subjects, for which the $O'$ state of the object is more favorable than the $O$ state. Description of qualitative characteristics (intention) shall involve the following symbols:
(+) means the subject is ready for $O \rightarrow O'$ transition (positive intention)
(–) means the subject resists $O \rightarrow O'$ transition (negative intention).
(±) means the subject is indifferent to $O \rightarrow O'$ transition, which is the result of absence of any winning/loss of the subject after transition to $O'$, or confidence that it is impossible to transfer the object to that state (zero intention).
The subject's intention $x \in N$ shall be denoted as $(\cdot)_x$, the subject $x$'s perception of the subject $y$'s perception shall be denoted as $(\cdot)_{xy}$, etc. Therefore subject $x$ maybe denoted as a tree (fig.2) that may be considered as the awareness dot structure described in [2].

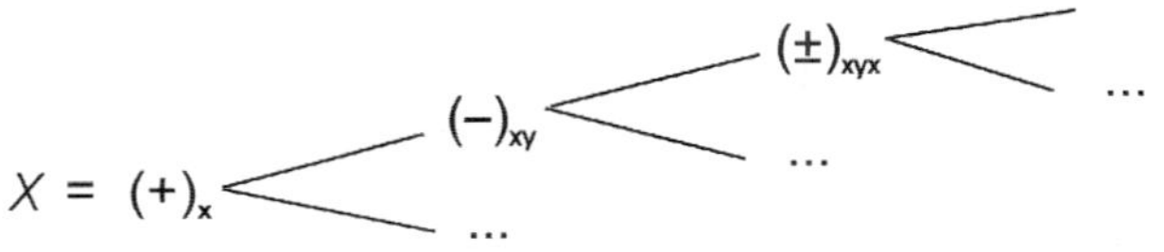


**Figure 2**. Model of the subject *x*.

Sequential notation of the tree's vertices $(+)_x$, $(-)_{xy}$, $(\pm)_{xyx}$, shall be referred to as "signature". As there is only one path to each of the vertices signatures shall be defined by the identifier of the last vertice, i.e. $(+ - \pm)_{xyx}$. In case general properties of signatures irrespective to specific subjects are researched then signature identifiers will be omitted.
The subject's readiness to take actions shall be denoted by putting the Boolean readiness function in accordance with the model's signature $f$: $(\cdot)_{x...} \rightarrow \{0,1\}$. Axioms introduced are $f(-)=1$, $f(+)=1$, $f(\pm)=0$. Readiness for other signatures is expressed by introducing rules (hypotheses) "rational conclusions", which agents use for making decisions on the possibility to take actions. These rules can be introduced in an agent's utility function matrix. In order to make the most beneficial action in terms of the utility function agents predict actions of counterparts. However as it is impossible to find solutions in pure strategies we suppose that choice of agents depends on the reflection level. The example below contains transitions beneficial for agents under the following scheme (fig.3).

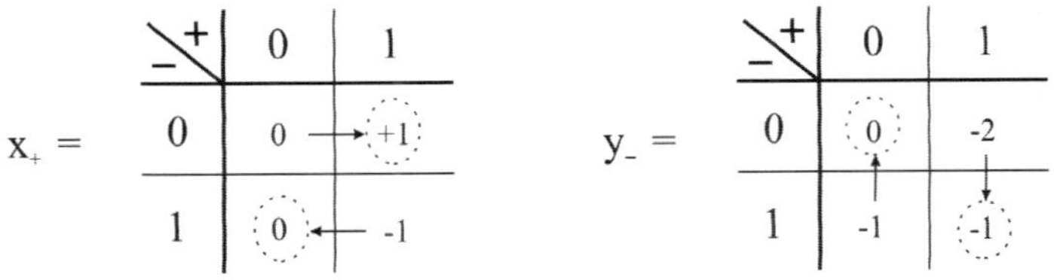


**Figure 3.** Transitions beneficial for agents.

In the matrix on the left when $y$ does not take actions (chooses 0 being the readiness value), hence action (1) is beneficial for $x$ and vice versa, i.e. $x$ benefits from applying the readiness value opposite to the value of $y$.

In the matrix on the right agent $y$ benefits from applying the same readiness as $x$.

In order to unify notation of such transitions S shall denote a set of all conceivable linear signatures, i.e. any sequences of elements $\{+, -, \pm\}$. As a result the rules that belong to either matrix for any of $\lambda$ and $\gamma$ contained by S may be represented as shown in fig.4.

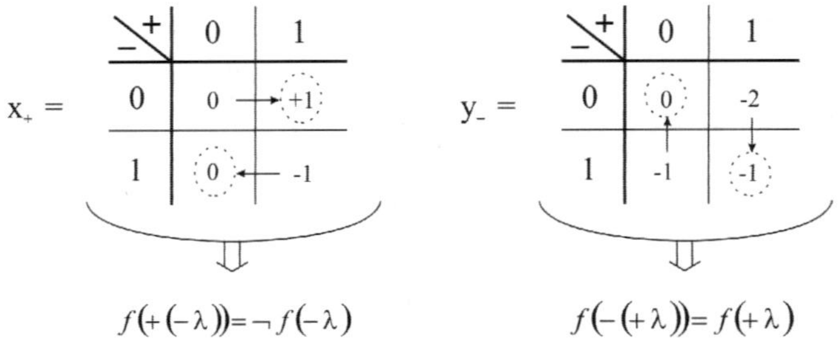


**Figure 4.** Accordance of transitions rules and utility matrix.

Thus rules express the dependency of element's readiness and the element following it in the awareness structure. Then let us generalize the situation and suppose that we can have a set of agents complying to the model above, which is expressed in an intention, readiness and reflexive structure.

The rules above may be regarded as first-order rules, i.e. ones created on the first branch of a tree (signature).

## 3. First-order rules

Rule 1.1. *The subject resists the transition with the same readiness as the counterpart is ready to perform it.*

Rule 1.2. *The subject does not resists the transition when there are other subjects resisting the transition and ready to act.*

1.1: $f(-(+\lambda))=f(+\lambda)$ (fig.4)

1.2: $f(-(-\lambda))=\neg f(-\lambda)$. (resource saving rule)

Rule 2.1. *The subject is ready to perform a transition only when the counterpart is not ready to take actions.*

Rule 2.2. *The subject does not perform a transition when there are subjects with a positive intention ready to take actions.*

2.1: $f(+(-\lambda))= \neg f(-\lambda)$ (fig.4)

2.2: $f(+(+\lambda))= \neg f(-\lambda)$. (resource saving rule)

Rule 3. *The subject with a negative intention, which does not observe the counterpart is not ready to take actions:*

3: $f(-(\pm\lambda))=0$.

Rule 4. *The subject with a positive intention which does not observe the counterpart is always ready to take actions:*

4: $f(+(\pm\lambda))=1$.

It is evident that $f(\pm(\lambda))=0$.

Based on the rules introduced it is possible to determine subject $x$'s readiness to take actions with $(+-+)_{xyx}$ :

$f(+)=1$ according to the axiom;
$f(-+)=f(+)=1$ according to Rule 1.1;
$f(+-+)=\neg f(-+)=0$ according to Rule 2.1.

Therefore the signature $(+-+)_{xyx}$ is a blocking one for *x*. The subject does not perform actions with such signature. A similar calculation states that $(+-+-+)_{xyxyx}=1$. The theorems below can be demonstrated:

***Theorem 1.***

$$f\left(\underbrace{+-}_{1}\underbrace{+-}_{2}\cdots\underbrace{+-}_{n}\sigma\right)=\begin{cases}1, \text{ when n is even,}\\ 0, \text{ when n is odd,}\end{cases}$$

σ equals either to (+) or to an empty set.
Many other theorems can be similarly introduced and proved, which associate a subject's set of signatures with readiness status.

## 4. Second-order rules

Second order rules are defined not in the linear signature but in two branches of the subject's awareness structure tree (fig.1).

<u>Rule 5.</u> *A subject with a negative intention takes actions only when the counterpart is ready for active actions and the other subject with a negative intention is not ready for active actions:*

5: $f(-(+\lambda)(-\gamma))=f(+\lambda)\wedge\neg f(-\gamma)$.

<u>Rule 6.</u> *The subject with a positive intention takes actions only when the counterpart is not ready for active actions and the other subject with a positive intention is not ready to take actions:*

6: $f(+(+\lambda)(-\gamma))=\neg f(+\lambda)\wedge\neg f(-\gamma)$.

<u>Rule 7.</u> *The subject with a negative intention observing the two counterparts takes actions only when one of the counterparts is ready to take actions:*

7: $f(+(-\lambda)(-\gamma))=\neg f(-\lambda)\wedge\neg f(-\gamma)$

<u>Rule 8.</u> *The subject with a negative intention observing the two counterparts takes actions only when one of the counterparts is ready to take actions:*

8: $f(-(+\lambda)(+\gamma))=f(+\lambda)\vee f(+\gamma)$

As a result a generalization can be made:

***Theorem 2.***

$$f\left(+\underbrace{(+(+\ldots(+}_{n})(-)\ldots\right)=1,$$ when n (number of brackets) is divisible by 3 and each nested signature complies to the constraints of $(+(+\lambda)(-\gamma))$ or $(-(+\lambda)(+\gamma))$.

Thus the agent with a positive intention observing the two other agents having different intentions and reckoning that they are not mistaken in their realizations is ready to take actions when its reflection level is divisible by 3.

Conclusion from theorem 2. When an agent does not have any preferences in choosing the reflection level then the probability that *x* will be ready to take actions for the latter example equals to 1/3 and for an example with two agents it's ½. Therefore an agent with a negative intention may introduce nonexistent agents into the real counterpart's perception and therefore decrease its readiness in order to decrease the risk of making a transition in the system.

***Theorem 3.*** $f(+(-)(-))^n = 0 \Leftrightarrow (\text{n - 1})$ divisible by 3 .

***Theorem 4.*** $f(-(+)(+))^n = 1 \Leftrightarrow (\text{n - 1})$ divisible by 3 .

***Theorem 5.*** $f(-(+)(-))^n = 0 \Leftrightarrow (\text{n - 1})$ divisible by 3 .

***Theorem 6.*** $f(+(+)(+))^n = 1 \Leftrightarrow$ n divisible by 2 .

**Conclusion**

The method of conflict situation analysis reviewed in the present paper is more convenient in some cases as it allows reviewing decisions concurrently at all possible agent reflection levels. However it significantly simplifies a conflict which is unacceptable in some cases. For instance in a situation with three agents we observe evident isomorphism for *k* agents, which is more likely not to be observed in reality and it resulted from a very rough description of agent types and rules of their decision making. However the model is quite appropriate for predicting behavior of a small number of agents without introduction of a utility function which is sometimes difficult to establish.

**References**

[1] Lefebvre V A 1999 Sketch of Reflexive Game Theory *Proceedings of the Workshop on Multi-Reflexive Models of Agent Behavior* (Los Alamos, NM) pp. 11-40.

[2] Lefebvre V A 1972 A formal Method of Investigating Reflective Processes. *General Systems, XVII,* pp. 181-188.

[3] Novikov D and Chkhartishvili A 2014 Reflexion and Control: Mathematical Models *London : CRC Press* 298 p.

[4] Zhu Q and Basar T 2013 Game-Theoretic Approach to Feedback-Driven Multi-Stage Moving Target Defense. *4th International Conference on Decision and Game Theory for Security, GameSec*, pp 154-162

[5] Singh A and Lakhotia A 2011 Game-theoretic design of an information exchange model for detecting packed malware. *6th International Conference on Malicious and Unwanted Software*, Malware 2011; Fajardo; Puerto Rico

[6] Cavusoglu H, Mishra B and Raghunathan S 2004 A model for evaluating IT security investments. *Communications of the ACM* Volume 47, Issue 7, pp 87-92.

# PHASE METHODS FOR MEASURING THE SPATIAL ORIENTATION OF OBJECTS USING SATELLITE NAVIGATION EQUIPMENT*

**Yu L Fateev[1], D D Dmitriev[1], V N Tyapkin[2], I N Kartsan[2], A E Goncharov[2]**

[1] Associate Professor, Siberian Federal University (SFU), Krasnoyarsk, Russia
[2] Associate Professor, Siberian State Aerospace University (SibSAU), Krasnoyarsk, Russia

E-mail: kartsan2003@mail.ru

**Abstract.** This paper discusses phase methods for measuring the spatial orientation of objects using satellite navigation equipment. An analysis of methods for resolving phase ambiguity is presented in the paper. The effectiveness and applicability of the single-step methods are discussed in detail. The authors have shown that applying the exhaustive method requires a constellation of 5 – 6 visible satellites. The measurement of signals, having a base length of 1 m transmitted from 8 satellites, demonstrates that an unambiguous solution is achieved practically in all cases.

**1. Introduction.** Professional development of navigational equipment in terms of increased accuracy is associated with the application of phase measurement methods in navigation signals [1]. Their application in survey-grade equipment enables determining the relative coordinates of an object within an accuracy of a few millimeters. Another application of phase methods is the determination of the spatial orientation of objects within GLONASS / GPS signal range.

The angular position of an object in space using satellite signals can be determined by measuring the difference in the path of signals between antennas located at the ends of vector bases. In order to improve the accuracy of spatial orientation, interferometers having a distance of several meters between the antennas (base length) are used. The main problem of phase measurements is phase ambiguity; this is a result of the short wavelength of the measured signals – about 19 cm. This signal is much less than the length of the baselines of the interferometer.

Currently, the majority of angle measuring and geodetic equipment for resolving ambiguities applies the LAMBDA method. This method requires at least two measurements considering that the phase ambiguity of the measurements does not vary. Thus, single-step methods for resolving ambiguity are of particular interest. Single-step methods based on the

* The research was done with the financial support of the Ministry of Education and Science of the Russian Federation in accordance with the agreement from 28.11.2014 № 14.577.21.0155, unique ID project RFMEFI57714X0155.

maximum likelihood use the redundancy of the system of equations; they can be obtained through the excess constellation of navigation satellites.

In a single-step method for resolving ambiguities an exhaustive search method is used in the interferometer. The solution is selected through the maximum likelihood. The constellation for implementing the exhaustive search method must contain at least 5 – 6 observable navigation satellites. When measuring signals from 8 navigation satellites with a base length of 1 m, almost all cases have a unique solution.

Modern satellite radio-navigation systems can determine the actual coordinates of an object within an accuracy of 3 – 5 m and a velocity vector within an accuracy of 0.1 m/s. Time-frequency binding equipment on the basis of the GLONASS / GPS radio-navigation systems is widely applied, enabling the synchronization of the onboard time scale with the Universal Time Coordinated (UTC) scale within an accuracy of 100 ns.

Further development of radio-navigation equipment is directly related to the application of phase methods for measuring parameters of navigation signals. The application of phase methods in geodetic equipment enables to determine the relative coordinates of an object within centimeter and millimeter accuracy.

Another application of phase methods consists in determining the spatial orientation of an object using GLONASS/GPS signals. Unlike magnetic sensors and inertial systems for measuring the spatial orientation of an object, satellite goniometric equipment determines the angles of course, pitch, and roll with respect to the true meridian – in this case the measured parameters are free of drift. Additionally, satellite goniometric equipment requires less initialization time and in comparison to inertial sensors is more cost efficient.

SRNS signals can determine the angular position of an object in space by measuring the difference of the navigation satellite signal course between the antennas situated at the endpoints of base-vectors. To determine spatial orientation only two non-collinear base-vectors (three antennas) are required.

The relationship between the phase shift of a navigation satellite signal, received by two spaced antennas and the cosine of the angle between the base-vector and the vector directed towards the navigation satellite (Fig. 1) is

$$\cos\alpha = \frac{\lambda \cdot \varphi}{2\pi B} \tag{1}$$

where $\lambda$ is the wave length of the satellite signal, $\varphi$ is the phase shift, $B$ is the base length, $\alpha$ is the angle between the base-vector and the vector directed towards the satellite. This expression is a single-base interferometer equation, which is widely applied in the theory of phase position-finders and antenna arrays.

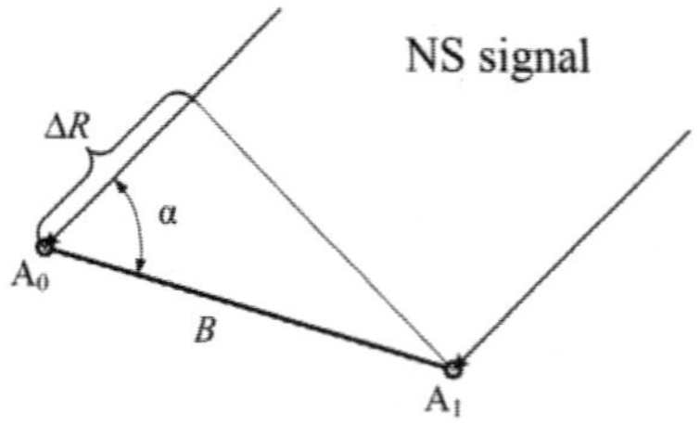


**Figure 1.** Single-base interferometer.

The coordinates of a base-vector can be determined from an equation on the basis of the scalar product of vectors:

$$k_x x + k_y y + k_z z = \Delta R = \lambda \frac{\Phi}{2\pi} \tag{2}$$

where $k_x, k_y, k_z$ are the direction cosines of the vector directed towards a navigation satellite, $x, y, z$ are base-vector coordinates, $\Delta R$ is path difference, $\Phi$ is phase shift of signals, $\lambda$ is wave length.

At least three equations are required to determine all unknowns. Considering that the coordinates of the base-vector are interrelated and having an acquainted base length $B$ the system of equations is

$$\begin{cases} k_{x,j} x + k_{y,j} y + k_{z,j} z = \Phi_j, \\ x^2 + y^2 + z^2 = B^2 \end{cases} \tag{3}$$

where $j = 1, \ldots, N$ is the visual satellite ordinal number.

Phase ambiguity is a major issue in phase measurements. Interferometers with a distance between antennas (the base length) exceeding several meters are used to improve the accuracy of determining spatial orientation. The ambiguity of phase shift measurement is caused by the fact that the wave length of the measured signals is rather small – approximately 19 cm. This is significantly smaller than the length of interferometer bases.

Methods for resolving phase ambiguity can be subdivided into two categories – single-step methods (apply the results of each measurement [2, 3]) and methods based on filtration (require the measurements of phase shifts during some time interval [4, 5])

$$k_x x + k_y y + k_z z = \Phi_j + n_j \lambda_j \tag{4}$$

where $n_j$ is the integer phase ambiguity.

Currently, most goniometric and geodetic equipment utilizes the LAMBDA method for resolving phase ambiguity [6, 7]. At the first stage, according to this method, the integer ambiguity $n$ is assumed to be an additional unknown without considering its integer value. Then each measurement gives one unknown $n_j$. In result of each separate measurement for $N$ navigational satellites, the system of equations will include $N+3$ unknowns, which requires more equations. Therefore the LAMBDA method requires at least two measurements, considering that phase ambiguity does not vary when measured. However, because the position of the satellite changes slowly, having a motionless object for each new measurement in (2) intensely correlates them with the previous ones. The system of equations degenerates, despite its redundancy. Decorrelation is applied to solve the ill-condition system of equations – this enables to assess part of the unknowns $n_j$ with a certain degree of accuracy. On the second stage, the obtained values of phase ambiguity are reduced to the integer form (generally by rounding); further, the initial system of equations is solved.

**2. Single-step method for resolving phase ambiguity.** Single-step methods for resolving phase ambiguity [8-10] are of specific interest. Based on the maximum likelihood, these methods utilize the redundancy from the system of equations, which can be obtained by using redundant navigation satellite constellations.

The single-step method utilizes the exhaustive method for phase ambiguity resolution in a single-base interferometer. A solution is selected, using the maximum likelihood criterion. When receiving signals of $N$ navigation satellites, the likelihood function (LF) is

$$W\left(\Phi_1,\Phi_2,\ldots,\Phi_N \mid x,y,z\right)=\prod_{j=1}^{N}\left[\frac{1}{\sigma_j\sqrt{2\pi}}\right] \cdot\exp\left[-\sum_{j=1}^{N}\frac{\left[\Phi_j+n_j\lambda_j-\left(k_{x,j}x+k_{y,j}y+k_{z,j}z\right)\right]^2}{2\sigma_j^2}\right] \quad (5)$$

with the additional condition

$$x^2+y^2+z^2=B^2 \quad (6)$$

Function (5) possesses local minima for every combination of ambiguities $n_j$. In order to minimize the likelihood function with respect to all possible values of $n_j$ an exhaustive search is conducted. The main disadvantage of this method is the large number of combinations of ambiguities $n_j$. When receiving signals from $N$ satellites, the number of combinations of ambiguities is $n_{\max}^N=\mathrm{int}\,(2B/\lambda+1)$ where $\mathrm{int}(s)$ is the integer part of a real $s$. For example, having a base length $B=1$ m the ambiguity $n_j$ for each navigation satellite may have 11 values (from 5 to 5). The total number of combinations of ambiguities is $11^3 = 1331$ for three navigation satellite measurements; $11^4 = 14641$ for four satellites; and $\approx 2\cdot10^8$ for eight satellites. A local minimum of (5) corresponds to each combination of $n_j$. For a large number of $n_j$ combinations, the values of local minima approach a global minimum. This increases the risk of an error. To reduce the number of calculations for resolving phase ambiguity, the base length can be decreased. However, this results in a poor accuracy of angular measurements.

The number of calculations may be considerably reduced provided that an initial constellation with a minimal number of satellites – a nonredundant constellation – is selected. Analyzing all possible combinations of phase ambiguity and solving the problem of phase ambiguity for these values, an initial set of solutions is obtained. Further each solution from the initial set is verified within the entire constellation. Additionally, false solutions are rejected by the maximum likelihood criterion or, equally, by an acceptable total disparity for a solution of the least squares minimum (LSM).

The likelihood function can be used to analyze the potential of the exhaustive methods. The angular position of a base-vector for a known length can be defined by two parameters – the course angle $K$ and the pitch angle $\Psi$. Hence, the likelihood function is two-dimensional. The course and pitch angles are related with the rectangular coordinates through the following expressions

$$X = B \cdot \cos \mathrm{K} \cdot \cos \Psi ;\ Y = B \cdot \sin \mathrm{K} \cdot \cos \Psi ;\ Z = B \cdot \sin \Psi . \tag{7}$$

When resolving phase ambiguity, the probability of gross errors – cases where phase ambiguity is determined with errors – is of special interest. Gross errors arise when the likelihood function has subordinate maxima which are comparable in magnitude with the main maximum, which corresponds to a correct solution. In Fig. 2 the likelihood function is given for one satellite. It isn't possible to resolve phase ambiguity with a separate measurement for one base of each navigation satellite because the likelihood function has extremums in whole areas, making errors indistinguishable from correct solutions.

Increasing the number of visible satellites transforms the total disparity to a sum of wave-like functions, obtained through measurements for each satellite and being a result of their interference. In Fig.3 we can see the likelihood function for four visible satellites. The main and subordinate maxima can be clearly distinguished.

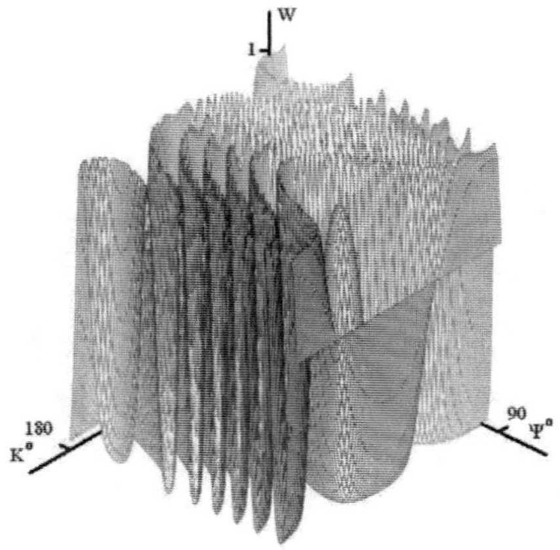


**Figure 2.** LF with measurements for one navigation satellite.

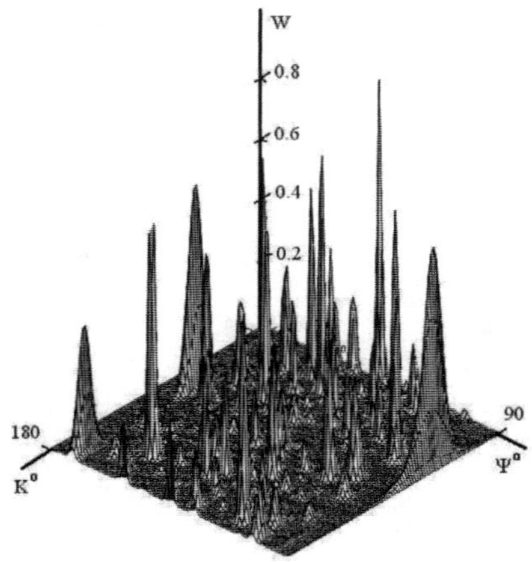


**Figure 3.** LF with measurements for four navigation satellites.

Since the likelihood function is complicated for analyzing there is demand for a parameter that enables estimating the probability of not making a correct solution and the probability of gross errors, i.e., the probability of making a false solution. The index of the likelihood function can be taken for this parameter, being the total residual of the solution of the LSM, which is equal to the sum of squares of residuals for all satellites or the square root of this value.

The residuals have two components: the result of residuals in subordinate maxima after a false solution and the result of dispersion in the measured phase shifts. A subordinate maximum of the LF provides a false solution. This makes the system of equations incompatible even for zero values of phase shift measurement errors in case of redundancy. In the absence of measurement noise, the magnitude of a residual in a subordinate maximum of the LF depends on the shape of the satellite and on the values of phase ambiguity. Hence, this value can be accepted as the mathematical expectation of residuals. The noise error of the phase shift measurements has a normal distribution. Thus, when solving (3), residuals for each satellite in the main and subordinate maxima of the LF are distributed by the normal law with mathematical expectation equal to residuals arising in the absence of measurement noise.

We shall consider the probability distribution function for the total residual. Provided that the mathematical expectations of the quantities $x_j$ are zero and their dispersions are equal to one another, the quantity $z = x_1^2 + x_2^2 + ... + x_N^2$ is distributed by the $\chi^2$ law with $n$ degrees of shift [10]. This takes place in the main maximum of the LF for the phase shift measurements with the same accuracy. Mathematical expectation in a subordinate maximum is not equal to zero and the $\chi^2$ distribution law in the classical form cannot be applied.

The distribution function for the total residual can be obtained as follows. First, the probability density for the square of one random variable is obtained. Then, using addition rules for sums of random variables, the necessary probability density is obtained. Characteristic functions are used to calculate the distribution function for the square of the total residual [12].

The characteristic function of a squared random variable with nonzero mathematical expectation is

$$\Theta(v)=\frac{1}{\sqrt{1-2i\sigma^2 v}}\cdot\exp\left(\frac{im^2 v}{1-2i\sigma^2 v}\right). \tag{8}$$

The characteristic function of the sum of squares of independent normal random variables with nonzero averages is equal to the product of characteristic functions

$$\Theta_n(v)=\left(1-2i\sigma^2 v\right)^{-\frac{n}{2}}\cdot\exp\left(\frac{iv}{1-2i\sigma^2 v}\sum_{k=1}^{n}m_k^2\right). \tag{9}$$

Equation (9) implies a property of the characteristic function: it depends not on mathematical expectations of initial random variables but on the sum of their squares $m^2=\sum_k m_k^2$. The distribution function for the sum of the squares of residuals must have the same property.

The probability density can be obtained by the inverse Fourier transform of the characteristic function (9)

$$P_n(x)=\frac{1}{2\pi}\cdot\int_{-\infty}^{\infty}\Theta_n(v)\cdot e^{-ivx}dv$$
$$=\frac{1}{2\pi}\cdot\int_{-\infty}^{\infty}\left(1-2i\sigma^2 v\right)^{-n/2}\cdot\exp\left(\frac{ivm^2}{1-2i\sigma^2 v}\right)\cdot e^{-ivx}dv \tag{10}$$

The graphs of probability density for different values of $m$ for five observed satellites can be observed in Fig. 4. The graph proves that a false solutions for $m>5\sigma$ can be rejected.

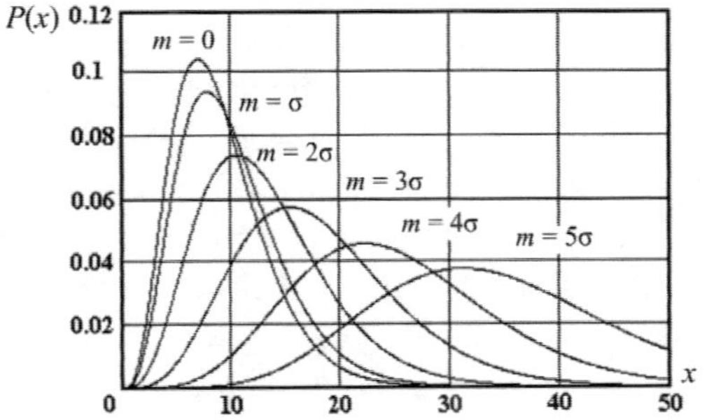


**Figure 4.** Probability density for the total residual for different values of $m$.

The probability density (10) cannot be expressed by elementary or tabulated functions making it complicated to calculate the probability of a given quantity in a specific area. An integral distribution function can be applied for this purpose

$$W_n(x)=\frac{1}{2\pi}\int_{-\infty}^{x}\int_{-\infty}^{\infty}\Theta(v)\cdot e^{-ivx}dvdx=\frac{1}{2\pi}\int_{-\infty}^{\infty}\frac{\Theta(v)}{iv}\cdot\left(1-e^{-ivx}\right)dv=$$
$$=\frac{1}{2\pi}\int_{-\infty}^{\infty}\frac{\left(1-2i\sigma^2 v\right)^{-n/2}}{iv}\cdot\exp\left(\frac{im^2 v}{1-2i\sigma^2 v}\right)\left(1-e^{-ivx}\right)dv. \tag{11}$$

In order to determine the probability of a false solution, given the probability of accepting a correct solution, it is necessary to determine the threshold value for which a correct solution is placed on the list of possible solutions with a given probability. The threshold value can be selected using the integral distribution function (11) with $m = 0$.

Fig. 5 illustrates the dependence of the probability of making a false solution on the ratio of the total mathematical expectation of residuals to their mean-square deviation (MSD) for different numbers of observed satellites. The probability of a false solution is characterized by a minimal mathematical expectation of the total residual in subordinate maxima. The efficiency of rejecting a false solution is achieved for $m > (5 \cdots 6)\sigma$.

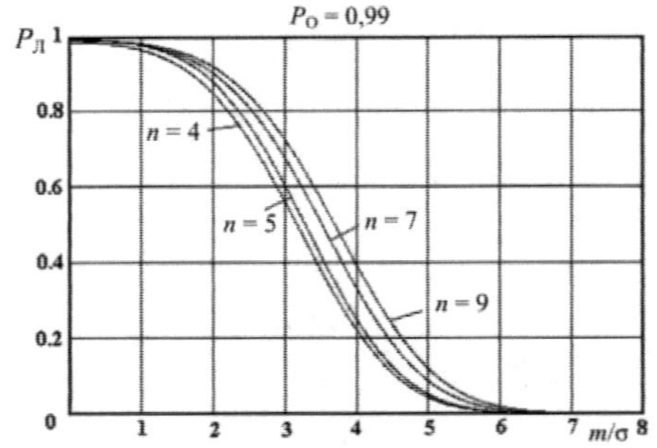


**Figure 5.** Dependence of probability of a false solution on the ratio of the total mathematical expectation of residuals to their MSD.

An exhaustive search of possible solutions results in a set of residuals produced by incorrect phase ambiguity resolution. The components are deterministic variables and represent mathematical expectation $m$. of residuals. They can be calculated a priori for each combination of phase ambiguities. The minimum value of these quantities is of special interest since in this case probability of a false solution is maximal and with increasing $m$ probability of a false solution decreases rapidly. However, the calculation of residuals for each specific case presents considerable difficulties, mainly because of a large number of combinations of phase ambiguity which take place in exhaustive search of all alternatives.

In the analysis mathematical expectations of residuals in subordinate maxima (for zero errors of the phase shift measurements) can be considered as random variables. According to the obtained data, the distribution of mathematical expectations of residuals does not depend on the shape of the constellation of navigation satellites, a space position, and the length of the base-vector. However, the minimal value of the total residual decreases with increasing the length of base. This dependence is explained by a quadratic increase of the number of possible positions of a base-vector with increasing its length. The square root of the sum of squares of residuals (the total residual) is reasonably described by the normal distribution and the mean-square deviation does not depend on the number of satellites in a constellation and equals 28 mm. The considered number of satellites in a constellation varies from 4 to 13 for different positions and lengths of base-vectors. The case with the measurements for 4 satellites is an exception. Distributions for $n = 4$ and $n = 9$ are shown in Fig. 6. The normal distribution for a large number of satellites can be explained by a consequence of the central limit theorem. If the number of satellites exceeds 5, the average value of the total residual depends linearly on the number of satellites in a constellation.

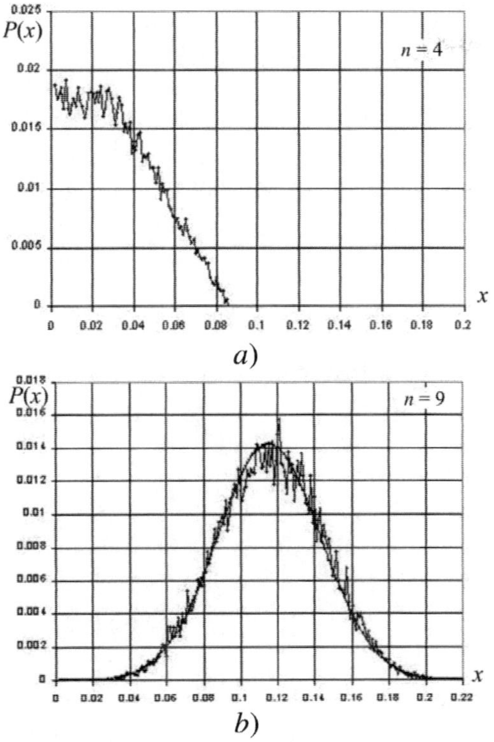


**Figure 6.** Distribution of the total residual for four (*a*) and nine (*b*) observed satellites.

The probability of making a false solution is mainly defined by the minimal residual in subordinate maxima. Using (11) for the integral distribution function we can determine the probability of in incorrect solution. In Figs. 7 and 8 the probabilities of gross error depending on the error of the phase shift measurements are illustrated for the minimal total residual with a base length of 1 m and 10 m.

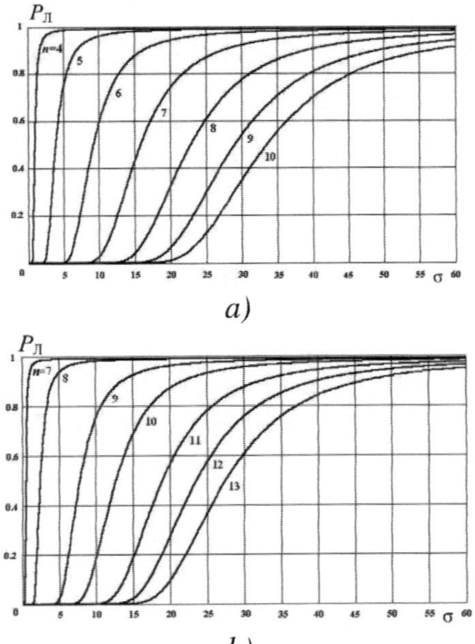


**Figure 7.** Probability of gross error for a base length of (*a*) 1 m and (*b*) 10 m.

**3. Conclusion.** The following conclusions can be made from the research:

1. The efficiency of the exhaustion method for the phase ambiguity resolution depends on the number of the observed satellites and the base length. For the length base of 1 m the exhaustion method can work successfully for 5 observed satellites; the noise error of the phase shift measurements equal to 5°. However, in a base length of 10 m for the same error of the phase shift measurements – 7 to 8 observed satellites are required.

2. The exhaustion method for the phase ambiguity resolution can be applied for an interferometer base length of up to 3 m and the limit MSD error of the phase shift measurements from 15 to 20°.

3. The constellation for implementing the exhaustion method should consist of at least 5 – 6 observed satellites. In order to measure the signals of 8 satellites and for a base length of 1 m we apply a unique solution for most cases.

It should be noted that the single-step exhaustion method for one base is applied to form an initial set of solutions. Hence, the probability of rejecting of a correct solution (which is defined by the threshold value of the likelihood function) is an important aspect. The presence of errors in the initial set does not mean there is a gross error, provided that a correct solution is involved as well. Further rejections of false solutions can be carried out by the filtering solutions of an initial set or by using of a multi-base antenna system.

**References.**

[1] Daneshmand S, Sokhandan N, Zaeri-Amirani M, Lachapelle G 2014 Precese calibration of a GNSS antenna array for adaptive beamforming application *Sensors* **14** pp. 9669–9691.

[2] Fateev Yu L, Dmitriev D D, Tyapkin V N, Garin E N, Shaidurov V V 2014 The phase ambiguity resolution in the angle-measuring navigation equipment *AIP Conference Proceedings* vol.1611 pp. 12–14.

[3] Hatch R R Method and apparatus for precision attitude determination and kinematic positioning US Patent № 4963889.

[4] Aleschechkin A M, Chmih M K, Fateev Yu L The method of the angular orientation of the object on the navigation signals spacecraft (options) RU Patent 2122217 MKI 6G01S5/02.

[5] Fateev Yu L, Garin E N, Dmitriev D D 2012 Measurement of angular orientation of rotating object *Radio Engineering* **9** pp. 69-75 (in Russian).

[6] Giorgi G, Teunissen P J G, Verhagen S, Buist P J 2012 Improving the GNSS Attitude Ambiguity Success Rate with the Multivariate Constrained LAMBDA Method *Geodesy for Planet Earth. International Association of Geodesy Symposia* vol.136 pp. 941–948.

[7] Teunissen P J G 2007 A general multivariate formulation of the multi-antenna GNSS attitude determination problem *Artificial Satellites* vol.42 **2** pp. 97–113.

[8] Denisov V P 1995 Analysis of quasi-optimum algorithm of the ambiguity elimination in multiscale phase-measuring system *Journal of Communications Technology and Electronics* vol. 40 **4** pp. 591-597.

[9] Fateev Yu L 2002 The determination of angular attitude on the basis of global navigation satellite systems *Radio Engineering* **7** pp. 51–57 (in Russian).

[10] Sushkin I N 2014 High-precision phase measure ements in the satellite navigation equipment of the user *Advance of Modern Radio Electronics. Foreign Radio Electronics* **5** pp. 57–60 (in Russian).

[11] Tikhonov V I 1966 Statistical radiotechnics (Moscow, Russia: Soviet Radio) (in Russian).
[12] Levin B R 1969 Theoretical Foundations of Statistical Radio Engineering (Moscow, Russia: Soviet Radio, 1969)

# THE CONCEPT OF CREATION OF INFORMATION SYSTEM FOR ENVIRONMENTAL MONITORING BASED ON MODERN GIS-TECHNOLOGIES AND EARTH REMOTE SENSING DATA

**Yu P Yuronen[1], E A Yuronen[2], V V Ivanov[3], I V Kovalev[4], P V Zelenkov[5]**
[1] Deputy Dean, Scientific And Educational Center "Institute Of Space Researches And High Technologies", Siberian State Aerospace University, Krasnoyarsk, Russia
[2] Associate Professor, Department Of Management Information Systems, Siberian State Aerospace University, Krasnoyarsk, Russia
[3] Director, Center Of Space Monitoring, Siberian State Aerospace University, Krasnoyarsk, Russia
[4] Professor, Second Russian doctoral degree in computer sciences, Siberian State Aerospace University, Krasnoyarsk, Russia
[5] Associate Professor, Department Of Systems Analysis, Siberian State Aerospace University, Krasnoyarsk, Russia

E-mail: erdas@sibsau.ru

**Abstract.** In this article the creation concept of the center of expeditious supervision and reaction for the solution of problems of environmental monitoring and support of adoption of administrative decisions is considered. The authors prove need of creation of the similar center in the territory of Krasnoyarsk region as the object consolidating existing and planned systems of land supervision and system of remote sensing.
Keywords: GIS-technologies, Earth remote sensing, regional information system.

Now in the territory of Krasnoyarsk region there is no integrated formalized system of expeditious environmental monitoring of natural and anthropogenic complexes. There are only separated for various reasons (lack of interdepartmental interaction, inconsistence or ignorance of opportunities of the existing centers) systems of conditionally expeditious monitoring of an environment's state in some parameters, such as air and water pollution, a radiation situation, a condition of a forest cover, etc. Thus there is a necessity of consolidation of the existing centers of reception, storage and data processing, both the Earth remote sensing, and these land supervision, for creation of united available system of environmental monitoring of the territory capable in real time quickly to reflect the existing situation on all square of supervision.

This problem can be solved by creation of the integrated center of collecting, storage and information processing from sources of land supervision (contamination control stations) and materials of Earth remote sensing which allow extrapolate and compare data of land supervision with space spectrozonal pictures. In the subsequent, with introduction of these data in the geographic information systems (GIS) supporting modern cartographical informational and WEB technologies, there is an opportunity to make a natural environment status evaluation in the territory of Krasnoyarsk region in real time [2]. Eventually it allows to react quickly to arising negative impacts and to develop optimum administrative and managerial decisions at various levels of executive power of the region [1].

The ideology of creation of system of environmental monitoring on the basis of a small, separate and uneven network of land stations and materials of space shootings allows trace and control an ecological situation. The advantage of this ideology is that there is an opportunity to trace and control an ecological situation in the territory with a big area as small local sites with big environmental risks, and ecology of a region "an masse". Thus, it is possible to say about scalability of this system what to allow essentially reduce the costs of creation of a uniform network of stations of the land supervision which number if necessary can be increased.

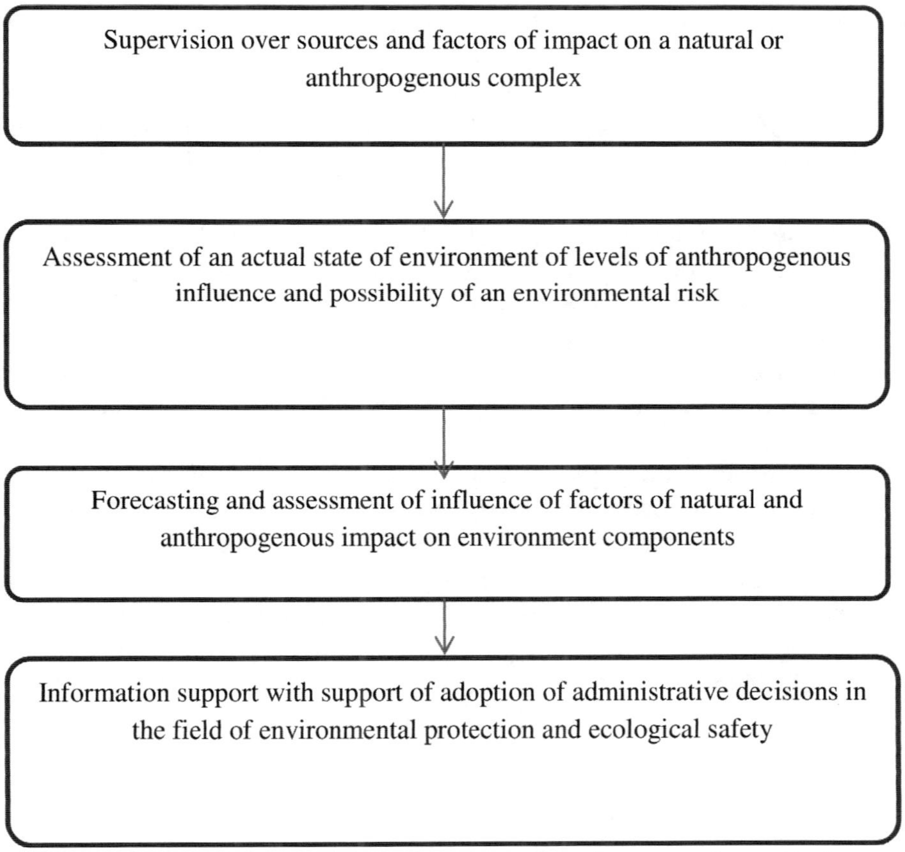


**Figure 1.** Scheme of creation of system of environmental monitoring.

In general the space pictures of Earth possess the following properties doing them by unique at expeditious environmental monitoring of territories:

- single multi-scale supervision of all territory;
- high degree of efficiency of obtaining information
- existence of continuous archives of space shooting;
- objectivity and reliability of received information;
- possibility of verification according to land supervision with the subsequent extrapolation on big squares;

- the automated generation of information products for the end user;
- various level of generalization and spectral characteristics;
- distributed access to databases.

These properties make space pictures by the irreplaceable tool at different solution of problems of an assessment of an ecological condition of all region or its local areas on which the risk of change of an ecological state is great.

The generalized function creation chart of such system can look as follows (fig. 1).

Having assumed as a basis the general scheme of creation of system of environmental monitoring on the basis of data of Earth remote sensing it is necessary to carry out the analysis of an existing state in this area in the territory of the region and to estimate needs for expeditious environmental monitoring. Further it is necessary to reveal a demanded data set and to develop the technological model of system allowing in an operational mode to carry out a complex assessment of an ecological condition of the region with the subsequent forecasting.

As a result of this work the following generalized scheme (fig. 2) which reflects as existing segments of planned system (the blocks allocated with green color), and those elements of system which demand creation or completion (the blocks allocated with blue color) will be received.

In general, the successful functioning of system first of all requires existence of data of Earth remote sensing. It is necessary to carry out retrofitting of existing stations of reception of satellite information to provide possibility of reception of data from satellites, both Russian, and foreign space segments (on the scheme are italicized the name of the satellites which are minimum demanded for functioning of system which aren't accepted now by any of the existing centers of reception of space information in the territory of Krasnoyarsk region).

The following step for the solution of a problem of creation of uniform system of environmental monitoring of edge is coordinating of activities of the existing centers of space monitoring of the Siberian federal university, the Siberian state aerospace university and the Ministry of Emergency Situations. It will allow to organize reserve systems of reception of space information and to relieve the centers of space monitoring of transmitted data duplication, and also to organize the coordinated uniform system of reception of given Earth remote sensing from numerous space group of satellites as being in operation, and only planned to a conclusion to an orbit.

IOP Conf. Series: Materials Science and Engineering **94** (2015) 012023 doi:10.1088/1757-899X/94/1/012023

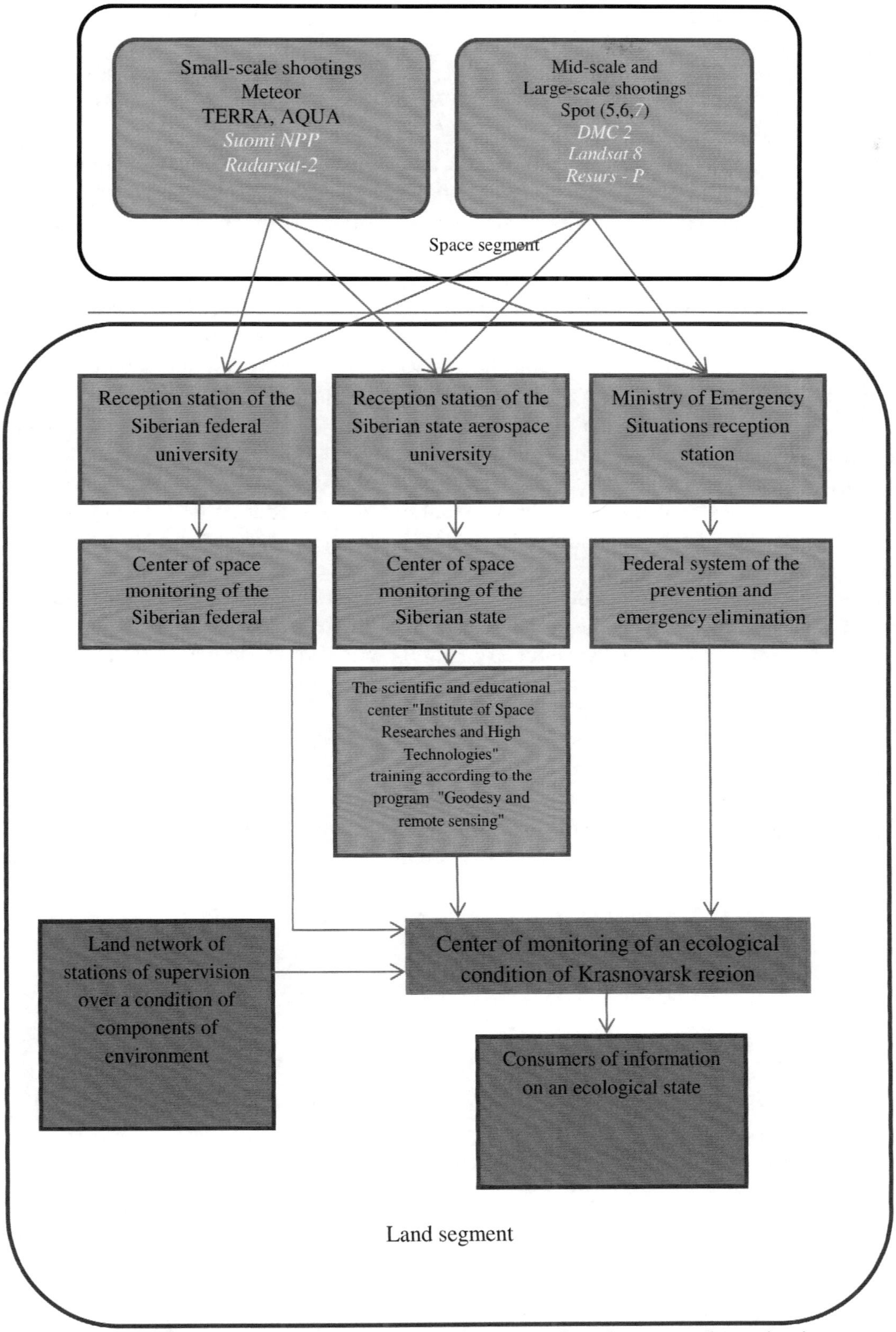


**Figure 2.** The generalized scheme of system of environmental monitoring.

Besides, productivity of functioning of the center of ecological monitoring and the organizations consumers of space information will depend in many respects on existence in the region of the specialized training center, capable to realize programs, both higher education, and short-term advanced training courses for specialists of various branch enterprises and executive authorities.

The center of environmental monitoring of the region with functions of coordinating, collecting, storage, data processing of space shooting and results of land supervision over a state of environment will be result of association of all these components. All obtained as a result of work of this center data can be informed to the consumer of information (both state, and private structures) by means of modern GIS in an operational mode.

Now it is necessary to consider in more detail structure of the center of monitoring of an ecological condition of Krasnoyarsk region (fig. 3). Apparently from this scheme the center can have branched system of the divisions having various functions and united by the general GIS and information system of collecting storage and processing of various data. [3]

All this allows system to function as a unit and to aggregate all production cycles in one program environment that minimizes temporary delays at the organization of operational system of monitoring of an ecological condition of the territory and to accelerate process of development of the administrative and managerial decision. In addition in system the system of public informing of the population and feedback for rapid response to public statements can be realized.

In summary it is possible to tell there is an urgent need in the organization of such system in the territory of the considered region. It is dictated both a geographical arrangement, and the area on which it is required to conduct expeditious monitoring. Besides, active social and economic development of the region and, as a result, probability of emergence of the environmental risks connected with development of industrial infrastructure, active mining and carrying out exploration in all territory of Krasnoyarsk region is of great importance. Also offered scheme in the basis assumes high extent of adaptation to arising tasks not only ecological character, it can include the regional monitoring systems founded on results of land supervision. Considering that fact that all system initially has to use modern GIS-technologies, there is a possibility of realization on this platform of information and analytical system for support of adoption of administratively administrative decisions at the level of the region.

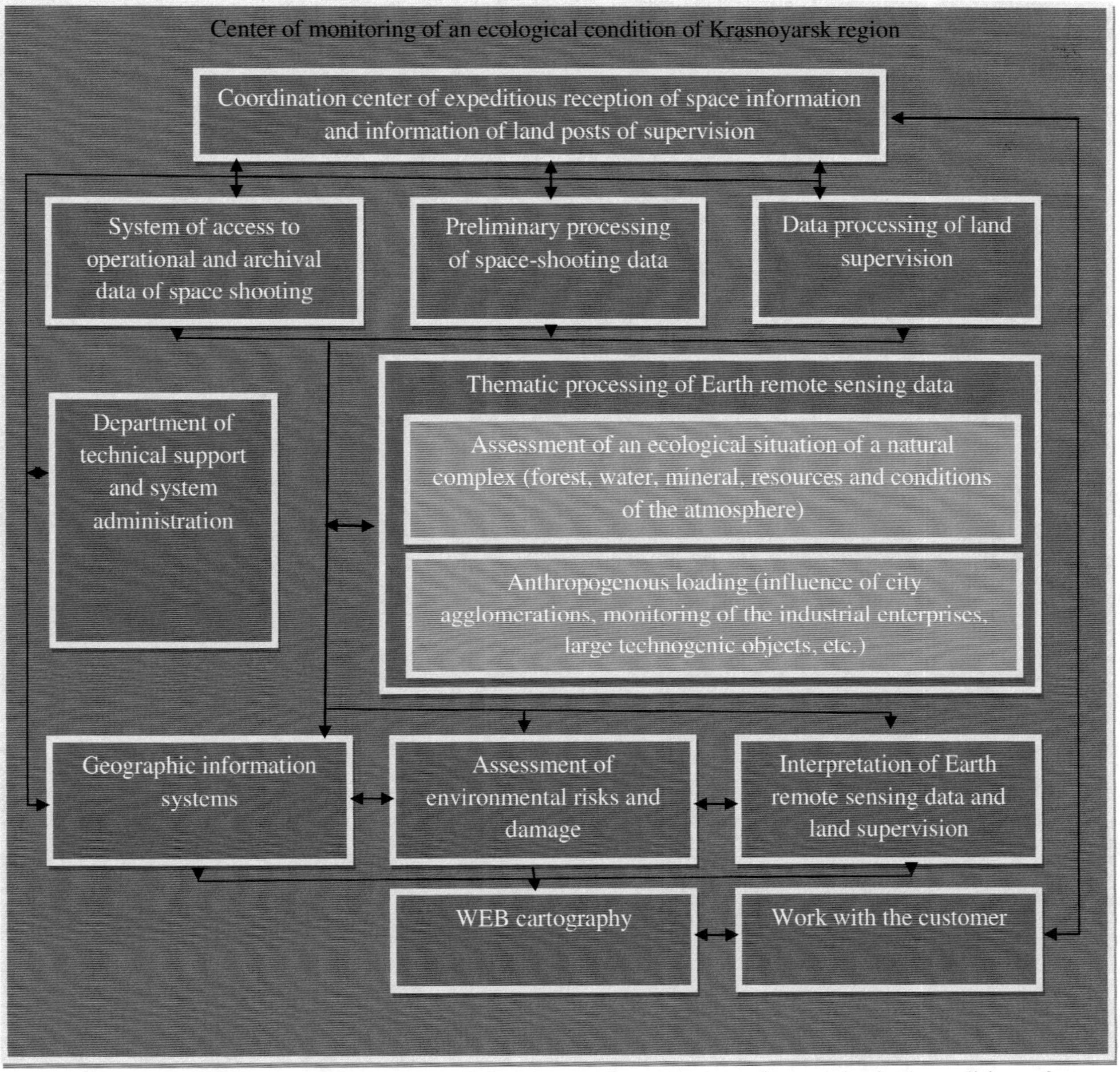


**Figure 3.** The integrated structure of the center of monitoring of an ecological condition of Krasnoyarsk region on carried-out functions of divisions.

**References.**

[1] Petrov M N, Yuronen Yu P 2007 *The inventory of natural resources of Krasnoyarsk territory - technology of control of forest users* Basic researches **2** 11.

[2] Loginov Yu Yu, Zelenkov P V, Yuronen Yu P, Ivanov V V, Borisevich A N 2013 *Development of the regional technological platform "Informational-telecommunication and space technology for innovative progress in Siberia"* Interexpo Geo-Siberia vol. 3 **4** pp. 159-165.

[3] Yuronen Yu P, Ivanov V V, Borisevich A N 2015 *Development of the operational ecological monitoring system of the Krasnoyarsk territory* Interexpo Geo-Siberia vol. 4 **2** pp. 98-102.

# Electron-hole pairs generation rate estimation irradiated by isotope Nickel-63 in silicone using GEANT4

I.V. Kovalev[1], V.G. Sidorov[2], P.V. Zelenkov[3], A.Y. Khoroshko[4], and A.T. Lelekov[5]

[1] Professor, Second Russian doctoral degree in computer sciences, Reshetnev Siberian State Aerospace University, Krasnoyarsk, RF

[2] Assoc. professor, First Russian doctoral degree in computer sciences, Reshetnev Siberian State Aerospace University, Krasnoyarsk, RF

[3] Assoc. professor, First Russian doctoral degree in computer sciences, Reshetnev Siberian State Aerospace University, Krasnoyarsk, RF
zelenkov@sibsau.ru

[4] Assoc. professor, First Russian doctoral degree in computer sciences, Reshetnev Siberian State Aerospace University, Krasnoyarsk, RF

[5] Assoc. professor, First Russian doctoral degree in computer sciences, Reshetnev Siberian State Aerospace University, Krasnoyarsk, RF
a.t.lelekov@yandex.ru

**Abstract.** To optimize parameters of beta-electrical converter of isotope Nickel-63 radiation, model of the distribution of EHP generation rate in semiconductor must be derived. By using Monte-Carlo methods in GEANT4 system with ultra-low energy electron physics models this distribution in silicon calculated and approximated with Gauss function. Maximal efficient isotope layer thickness and maximal energy efficiency of EHP generation were estimated.

## Introduction

New generation micropower devices and microelectromechanical systems need compact long-lifetime power supplies. Betavoltaic effect radioactive energy to electric energy transformation is one of perspective ways for meeting such requirements. Betavoltaic semiconductor converter of 63Ni isotope radiation is considered to be a perspective technology by multiple scientific researches in this area. 63Ni isotope emits beta-particles having the energy level below structural damage energy for most semiconductors, does not emit gamma-particles and has half-decay time at about 100 years.

Isotope undergoes radioactive decay

$$^{63}_{28}\mathrm{Ni} \to {}^{63}_{29}\mathrm{Cu} + \mathrm{e}^-,$$

the result of which is emission of electron ($\beta$-particle) having average energy 17.3 keV with a random direction. During travel through the space-charge region of p-i-n junction such particle causes the generation of electron-hole pair (EHP) which is being separated by the built-in electric field. Separated charge carriers travel to the opposite directions: holes into the p-region, electrons into the n-region. As the result, difference of potentials is induced on p and n region leads, the same way as it occurs inside photoelectric converter. If connectors are attached to external resistance, the current flows through the circuit and generated energy can be used to power the load.

The main advantages of radioisotope-based power sources are long service life (determined by half-decay time), high specific energy, high energy density, wide range of operational environment conditions (determined by semiconductor) and high reliability. It is possible to create flexible hybrid voltage or current source incorporating the charge accumulation for continuous operation or pulse-mode operation.

It is necessary to perform the modelling of all energetic processes – primary beta-electrons generation, electron-hole pairs generation, particles diffusion and drift inside space-charge region, recombination and others. It is possible to use COMSOL Multiphysics for modelling some of listed processes, but it doesn?t incorporate the model of betavoltaic particles generation. For this reason Monte-Carlo method modelling using GEANT4 software was performed to estimate the distribution of EHP generation rate in semiconductor.

## Problem definition

Different scientific research teams use different approach to estimate the the distribution of electron-hole pairs generation rate.

Betavoltaic effect modelling was performed in monograph [9]. It was proposed, that energy carriers generation rate is proportional to the distribution of energy generation rate through the depth of semiconductor. Generation rate for every particle energy (see fig. 1(a)) is estimated using table data from [5]. Energy generation rate was approximated using Gauss function

$$J(z) = A \exp\left(-\frac{(z-z_0)^2}{2\sigma}\right), \tag{1}$$

where parameters $z_0$ and $\sigma$ were extrapolated to the low-energy range.

Charge carriers generation rate dependency on electron penetration depth for $^{63}$Ni isotope with radioactivity 20 mKi is shown on figure 1(b). Charge carriers generation rate was calculated as

$$G = \int_0^{67\text{keV}} A(E) \cdot \Omega(E) \cdot J(E)\, \mathrm{d}E,$$

where $A(E)$ - amplitude of absorption function, $\Omega(E)$ - energy spectrum $^{63}$Ni isotope, $J(E)$ - energy generation function, estimated using equation (1). From graph of generation rate dependency on electron penetration depth one can estimate that maximum charge carriers generation rate located at the depth about 6 um.

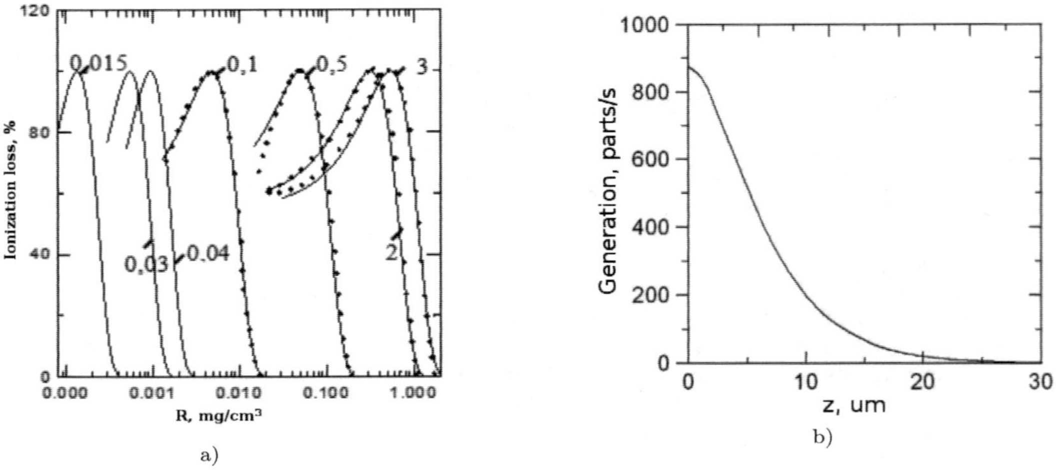


**Fig. 1.** a) Energy generation rate distribution by the depth of semiconductors for electrons having different energy. Numbers 0.015 - 3 corresponds to the radiation energy in MeV.
b) Charge carriers generation rate dependency of electrons penetration depth when using $^{63}$Ni, per one beta-particle. Graphs taken from [9].

The article [6] studies beta-particles energy loss in silicone. Calculation is performed using Monte-Carlo method. Plot graph of ionization energy losses for isotope $^{63}$Ni having mass density 1mg/cm$^2$ is shown on figure. The most of the energy loss takes place in near-surface layer of silicone, energy losses distribution is close to exponential. Total energy ionization losses is up to 80% at silicone thickness 5 um and near 100% at 20 um.

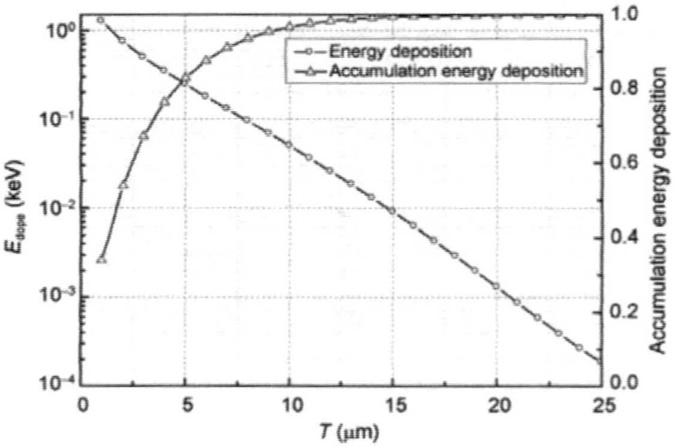


**Fig. 2.** Dependence of ionization losses of electron penetration depth when using $^{63}$Ni isotope (from [6]).

In the article [7] interaction of betaelectron radiation with silicone is calculated using Monte-Carlo method in CASINO software. The dependency of electron energy losses of depth of penetration is shown on fig 3. Authors observed that the energy loss curve for 10 keV particles has maximum at 400 nm, therefore p-n junction depth should be at least 500 nm to provide efficient operation.

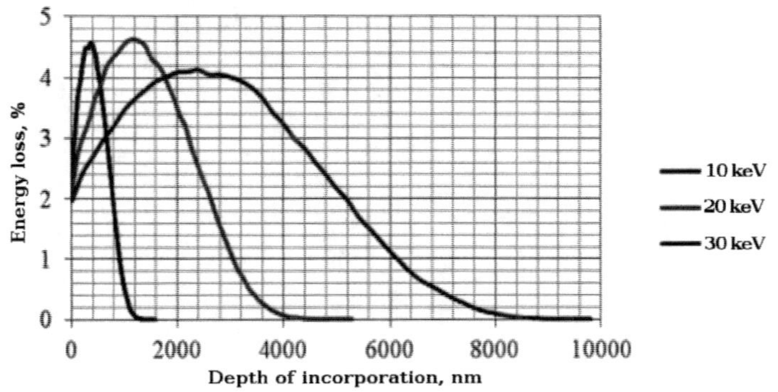


**Fig. 3.** Dependence of radiation energy ionization losses from electron penetration depth (from [7]).

The comparison of researches listed above shows that energy generation rate curves differ, which may be caused by simplification of electron-matter interaction models.

Current article propose to use the electron-matter interaction model, which is defined for low levels of energy. For instance, model proposed in [2] describe the ionization and electron transition into conduction zone for energy range from 16.6 Ev to 50 kEv. The calculated interaction cross-sections are used to plot the tracks of beta-electrons for isotope $^{63}$Ni in silicone.

The primary goal of current research is to build the model of electron-hole pairs generation rate in silicone. The selected geometrical schematic of this problem is shown on figure 4(a).

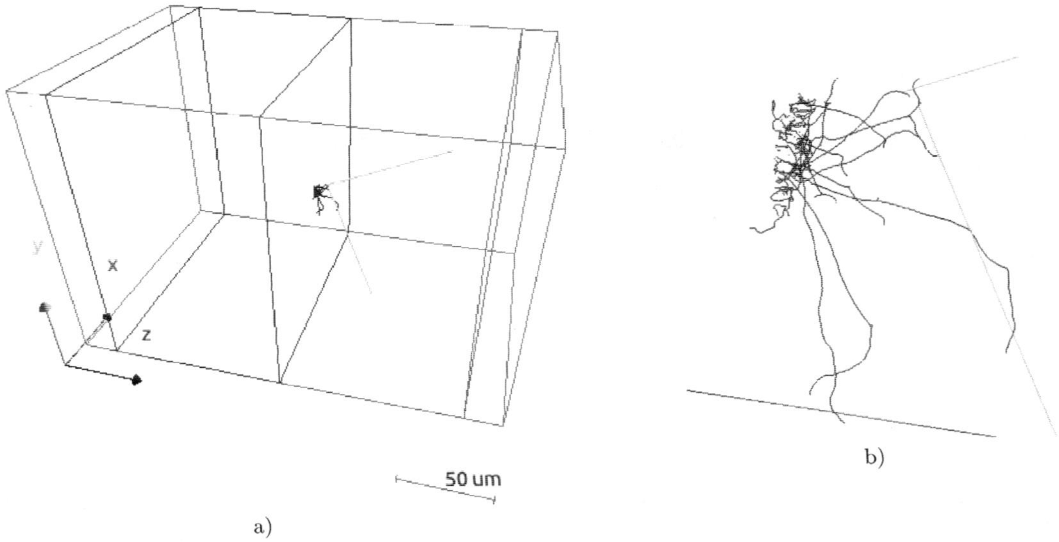


**Fig. 4.** a) geometry of the problem; b) the tracks of beta-electrons, emitted by $^{63}$Ni isotope (100 tracks).

The object is a cuboid consisting of two equal parts - isotope and silicone. Both parts are 400 um thick, which is equal to standard silicone wafer thickness. Particle tracks starting from the rectangle shaped area 6x6 um, placed at the plane between isotope and semiconductor. Figure 4(b) shows the resulting beta-electrons tracks in silicone.

Fast electrons generation takes place in the isotope layer. The spectrum of beta-radiation corresponds to $^{63}$Ni isotope (see fig. 5), spectrum is taken from article[1]. One can see that spectrum, approximated with exponential function $a\exp(-b\cdot E)$ differs from measured curve (criterion $R^2 = 0,991$). Despite this fact, exponential model was used in current research to increase computation speed, even though GEANT4 supports table-defined distributions.

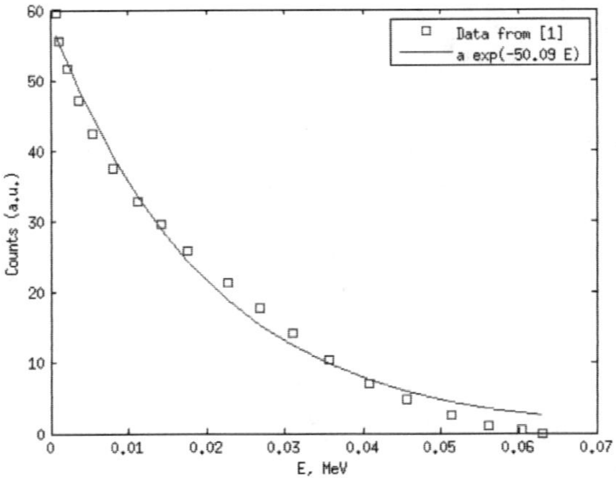


**Fig. 5.** $^{63}$Ni radioactivity spectrum approximation.

Calculations listed below do not take into account the built-in electric field of p-i-n junction space-charge region, despite the fact that it should influence the distribution of generation rate. Currently there is no dopant concentration distribution data for beta-voltaic converter being developed, for this reason space-charge region electric field estimation is not possible.

Calculation sequence:

1. Tracks calculation in GEANT4, corresponding to the software-defined experiment schematic;
2. Saving tracks data (nodal points coordinates, the amount and type of energy loss) as CSV-files;
3. Reading of generated CSV-files into memory, saving to float data files;
4. Separation the data of every track from whole data array, taking in consideration only ionization losses; secondary electrons are filtered out from tracks data, they are taken into account as separate particle tracks);
5. Ionization energy losses (the number of electron-hole pairs generated) is calculated for every elementary volume of silicone.
   Ionization energy losses calculation is performed using the following algorithm. For every track, the segment table of energy losses is filled, including starting and ending points of segment. The electron-hole pair energy in silicone is estimated as $E_{e-h} = 3,6$ eV.
   Using this estimation, the number and coordinates of generated electron-hole pairs is calculated, considering the probability of generation to be constant during the whole track segment. The space of semiconductor is divided into elementary volumes and for each one the quantity of generated pair is calculated.
6. Calculation of electron-hole pairs generation rate field in semiconductor.
   Estimated field of energy generation rate is normalized relative to total initial energy of beta-radiation, representative segment of field is separated and approximated using Gauss function [8]:

$$C(x,y) = \frac{N}{\sqrt{2\pi}\cdot R_P}\exp\left(-\frac{(y-R_P)^2}{2\,\Delta R_P{}^2}\right)\cdot\left(\mathrm{erf}\frac{x+a}{\sqrt{2}\,\Delta R_X} - \mathrm{erf}\frac{x-a}{\sqrt{2}\,\Delta R_X}\right), \tag{2}$$

   where $y$ – position on axis, perpendicular to semiconductor surface, um; $x$ – position on axis, parallel to surface; $N$ – injection doze, electron/cm$^2$; $R_P$ – projective range, um; $\Delta R_P$ – standard deviation of projective range, um; $a$ – beta-emitter width coefficient; $\Delta R_X$ – standard deviation of lateral range, um.
   First part of this equation was used to approximate one-dimensional (normal to surface, pointed into the depth of semiconductor) generation rate distribution (2):

$$C(y) = \frac{N}{\sqrt{2\pi}\cdot R_P}\exp\left(-\frac{(y-R_P)^2}{2\,\Delta R_P{}^2}\right). \tag{3}$$

GEANT4 source code used for track estimation is uploaded to GitHub[6]. Number of tracks used for simulation was 20000, to increase the simulation precision. Calculation was separated into 4 threads to minimize the simulation time.

[6] https://github.com/alex-khoroshko

## Results and discussion

The field of EHP generation rate in silicon after radiation influence of 3 um $^{63}$Ni layer and it's Gauss function approximation (2) is presented on figure 6. One-dimensional distribution (fig. 6, b) approximated with function $a \cdot \exp(-(y-b)^2/c^2)$, where $a = 7,184e+04$, $b1 = -68,74$, $c = 16,81$. One can see that the beta-particle flux generate EHP generally in depthward direction to silicon, to sideways the radiation disperse weaker. Most of EHP generated in subsurface layer, at depths not more 20 um.

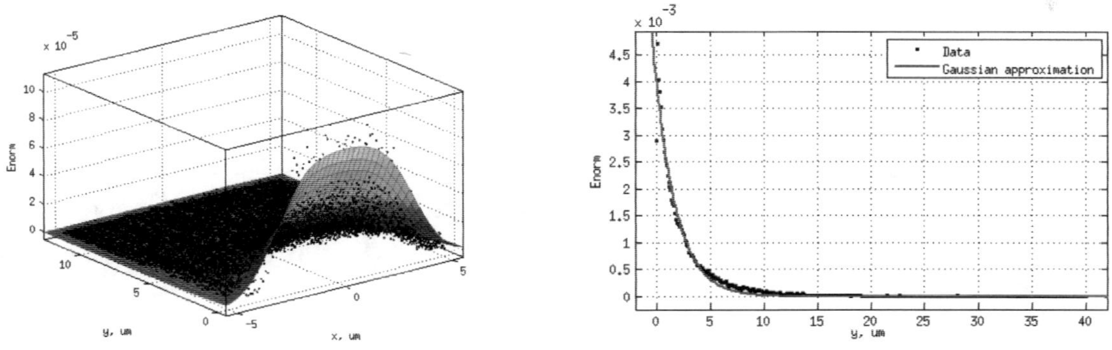


a) b)

**Fig. 6.** The distribution of EHP generation rate in the depth of silicone when radiated by $^{63}$Ni isotope (isotope thickness is 3 um, run 80000 tracks), a) two-dimensional, b) one-dimensional

To estimate of isotope layer thickness influence to EHP generation rate, spatial radiation source was replaced by plain surface (square 6x6 um, parallel to silicone surface). Radiating surface displaced from silicone surface to isotope region. All graphs are normalized relatively to initial particles radiation energy. Gauss function coefficient values are listed in the table 1.

**Table 1.** Coefficients of approximating function.

| | 2D, eq. (2) | | | | | 1D, eq. (3) | | |
|---|---|---|---|---|---|---|---|---|
| $\Delta y$, um | $N$ | $R_P$ | $\Delta R_P$ | $a$ | $\Delta R_X$ | $N$ | $R_P$ | $\Delta R_P$ |
| 0 | -2.2646e+08 | -161.3 | 48.518 | 3.004 | 0.24918 | -12153 | -67.2 | 6.5909 |
| 0.25 | -1.4792e+08 | -142.4 | 42.93 | 3.119 | 0.8358 | -1284.7 | -56.89 | 6.3953 |
| 0.5 | -1.0004e+08 | -147.1 | 44.331 | 3.194 | 1.0677 | -245.45 | -49.63 | 6.2145 |
| 1 | -6.2004e+07 | -127.9 | 38.299 | 3.095 | 1.2763 | -38.072 | -42.19 | 5.9699 |
| 1.5 | -2.666e+07 | -137.5 | 41.473 | 3.169 | 1.4093 | -3.4767 | -35.5 | 5.9447 |
| 2 | -1.3394e+07 | -141.7 | 42.729 | 3.639 | 1.7918 | -1.4363 | -30 | 5.5299 |
| 3 | -1.126e+06 | -131.7 | 41.11 | 3.12 | 1.8413 | -0.36347 | -26.15 | 5.246 |

Figure 7 shows the distribution of EHP generation rate along the normal to surface of silicone. The comparison of obtained distributions and distributions at figures 1 b) and 2, shows that distributions have same curve shapes, values match closely to each other.

Figure 8 shows the graph of total EHP energy depending on the distance of nickel particle-generating layer. Total emitted energy equals to 1, thus the graph corresponds to the energy efficiency of EHP generation. One can see that no more than 18% of initial energy is spent on EHP generation. This curve is approximated by the exponent $a \cdot \exp(-\Delta y/T_y)$, where $a = 0,1878$, $T_y = 0,7987\,\text{um}$. The layer having thickness of $3 \cdot T_y = 2,4\,\text{um}$ generates 95% of useful energy, further increment of thickness does not increase useful energy generation. Using this factor, one can estimate economically justified isotope thickness. Total generation efficiency for semi-infinite isotope layer can be estimated as

$$\eta = \int_L a \cdot \exp\left(-\frac{\Delta y}{T_y}\right)\, \mathrm{d}l = aT_y = 0,15.$$

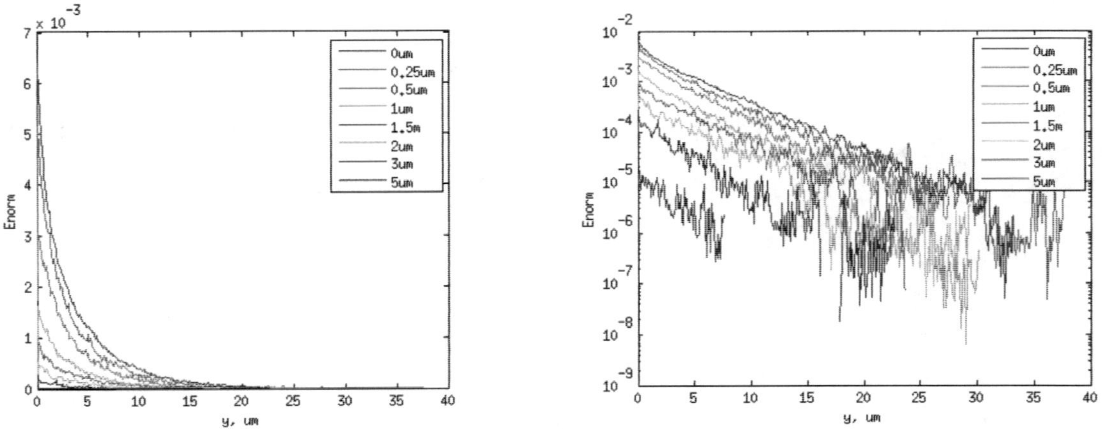


**Fig. 7.** The distribution of EHP generation rate with radiation source with different displacement.

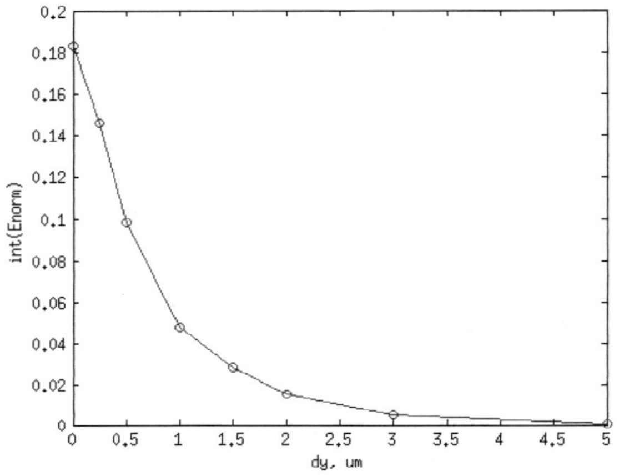


**Fig. 8.** Total energy

## Conclusion

As the result of modelling of EHP generation in silicone induced by $^{63}$Ni isotope radiation in GEANT4 with ultra-low energy models, maximal efficient isotope layer thickness was estimated - 2, 4 um. Maximal energy efficiency of EHP generation is about 15%. The fields of EHP generation rate with different thickness of $^{63}$Ni isotope were estimated, so that it is possible to optimize parameters of silicone semiconductor structure. Calculated Gauss function approximation parameters can be used to define the fields for numeric modelling of beta-electrical converters, in software such as Comsol Multiphisics.

Research is performed with support from Russian Federal Targeting Program «Research and development for priority targets of Russian scientific-technological complex development years 2014-2020», agreement 14.577.21.0117.

## References

1. Beta spectrometry with metallic magnetic calorimeters / M. Loidl, C. Le-Bret, M. Rodrigues, X. Mougeot // Applied Radiation and Isotopes (proceedings of the 19th International Conference on Radionuclide Metrology and its Applications 17-21 June 2013, Antwerp, Belgium) - 2014 - vol. 87 - pp. 302-305, doi:10.1016/j.apradiso.2013.11.024.
2. Geant4 physics processes for microdosimetry simulation: Very low energy electromagnetic models for electrons in silicon / A. Valentin, M. Raine, J.-E. Sauvestre, M. Gaillardin, P. Paillet. // Nuclear Instruments and Methods in Physics Research Section B: Beam Interactions with Materials and Atoms - 2012 - vol.288 - pp. 66 - 73, doi:10.1016/j.nimb.2012.07.028.
3. Geant4 User's Guide for Application Developers by Geant4 Collaboration. Version: geant4 10.1 // 2014. PDF file. URL: http://geant4.web.cern.ch/geant4/UserDocumentation/ UsersGuides/ForApplicationDeveloper /fo/BookForAppliDev.pdf (online; accessed: 05/10/2015).

4. Radiacionno-stimulirovanniy element pitaniya na osnove beta-istochnika / E.S. Pchelintseva, B.M. Kostishko // Ivestiya Samarskogo nauchnogo tsentra Rossiyskoy akademii nauk - 2012 - vol.14 - ?4 - p. 1059 - 1061. (in russian)
5. Spravochnik fizicheskih velichin / I.S. Grigorieva, E.Z. Meylihova. M.: Energoatomizdat, 1991. 1232 p. (in russian)
6. Optimization design and analysis of Si-$^{63}$Ni betavoltaic battery / XiaoBin Tang, Ding Ding, YunPeng Liu, Da Chen // SCIENCE CHINA. Technological Sciences. - 2012 - vol. 55(4) - pp. 990-996, doi:10.1007/s11431-012-4752-6.
7. Peculiarities of Betavoltaic Battery Based on Si / V.N. Murashev, V.N. Mordkovich, S.A. Legotin, O.I. Rabinovich, A.A. Krasnov // Journal of Nano- and Electronic Physics - 2014 - vol. 6(4) - pp. 04012(2cc).
8. Process and Device Simulation for MOS-VLSI Circuits / P. Antognetti, D.A. Antoniadi, R.W. Dutton, W.D. Oldham // Proceedings of the NATO Advanced Study Institute on Process and Device Simulation for MOS-VLSI Circuits, SOGESTA, Urbino, Italy, July 12-23, 1982, 619 p., ISBN-13: 978-94-009-6844-8, doi: 10.1007/978-94-009-6842-4.
9. Sovremennie aspekty primeneniya betavoltaicheskogo effekta / U.S. Nagornov // Ulyanovsk: UlGPU, 2012. 113 p. ISBN 978-5-86045-493-4. (in russian)

**Institute of Physics**
Dirac House, Temple Back
Bristol BS1 6BE UK

ISSN: 1757-8981
ISBN 978-1-5108-1465-3

CURRAN ASSOCIATES INC.
proceedings.com